Jens Dreyer

Neutrinos from Starburst-Galaxies

Jens Dreyer

Neutrinos from Starburst-Galaxies

A source stacking analysis with AMANDA-II and IceCube data

Südwestdeutscher Verlag für Hochschulschriften

Impressum/Imprint (nur für Deutschland/ only for Germany)
Bibliografische Information der Deutschen Nationalbibliothek: Die Deutsche Nationalbibliothek verzeichnet diese Publikation in der Deutschen Nationalbibliografie; detaillierte bibliografische Daten sind im Internet über http://dnb.d-nb.de abrufbar.

Alle in diesem Buch genannten Marken und Produktnamen unterliegen warenzeichen-, marken- oder patentrechtlichem Schutz bzw. sind Warenzeichen oder eingetragene Warenzeichen der jeweiligen Inhaber. Die Wiedergabe von Marken, Produktnamen, Gebrauchsnamen, Handelsnamen, Warenbezeichnungen u.s.w. in diesem Werk berechtigt auch ohne besondere Kennzeichnung nicht zu der Annahme, dass solche Namen im Sinne der Warenzeichen- und Markenschutzgesetzgebung als frei zu betrachten wären und daher von jedermann benutzt werden dürften.

Verlag: Südwestdeutscher Verlag für Hochschulschriften Aktiengesellschaft & Co. KG
Dudweiler Landstr. 99, 66123 Saarbrücken, Deutschland
Telefon +49 681 37 20 271-1, Telefax +49 681 37 20 271-0
Email: info@svh-verlag.de
Zugl.: Dortmund, TU, Diss., 2009

Herstellung in Deutschland:
Schaltungsdienst Lange o.H.G., Berlin
Books on Demand GmbH, Norderstedt
Reha GmbH, Saarbrücken
Amazon Distribution GmbH, Leipzig
ISBN: 978-3-8381-1762-1

Imprint (only for USA, GB)
Bibliographic information published by the Deutsche Nationalbibliothek: The Deutsche Nationalbibliothek lists this publication in the Deutsche Nationalbibliografie; detailed bibliographic data are available in the Internet at http://dnb.d-nb.de.

Any brand names and product names mentioned in this book are subject to trademark, brand or patent protection and are trademarks or registered trademarks of their respective holders. The use of brand names, product names, common names, trade names, product descriptions etc. even without a particular marking in this works is in no way to be construed to mean that such names may be regarded as unrestricted in respect of trademark and brand protection legislation and could thus be used by anyone.

Publisher: Südwestdeutscher Verlag für Hochschulschriften Aktiengesellschaft & Co. KG
Dudweiler Landstr. 99, 66123 Saarbrücken, Germany
Phone +49 681 37 20 271-1, Fax +49 681 37 20 271-0
Email: info@svh-verlag.de

Printed in the U.S.A.
Printed in the U.K. by (see last page)
ISBN: 978-3-8381-1762-1

Copyright © 2010 by the author and Südwestdeutscher Verlag für Hochschulschriften Aktiengesellschaft & Co. KG and licensors
All rights reserved. Saarbrücken 2010

Contents

1	**Introduction**	**1**
2	**Status of astroparticle physics**	**3**
	2.1 Brief historical development	3
	2.2 The energy spectrum of cosmic rays	4
	2.3 Particle acceleration	5
	2.4 Possible sources of cosmic rays	6
	2.4.1 Supernovae and Supernova Remnants	8
	2.4.2 Pulsars and binary systems	8
	2.4.3 Active galactic nuclei	9
	2.4.4 Gamma ray bursts	12
	2.5 Messenger particles and their detection	13
	2.5.1 Protons and heavy nuclei	13
	2.5.2 Photons	18
	2.5.3 Neutrinos from astrophysical sources	21
	2.5.4 Neutrino detection	22
	2.6 Astrophysical neutrinos	24
	2.6.1 AGN as neutrino sources	26
	2.6.2 Neutrinos from GRBs	27
	2.6.3 Neutrinos from Starburst-Galaxies	29
3	**Starburst-Galaxies**	**31**
	3.1 A local sample of Starburst-Galaxies	33
	3.1.1 FIR luminosity versus Radio power	35
	3.1.2 Infrared to radio flux density ratio	37
	3.1.3 Radio to Infrared and X-ray to Infrared spectral indices	38
	3.2 Starburst-Galaxies as neutrino sources	39
	3.2.1 Gamma Ray Bursts and starbursts	40
	3.3 Enhanced neutrino flux from GRBs in starbursts	42
4	**Neutrino telescopes at the South Pole**	**47**
	4.1 The AMANDA telescope	47
	4.2 The IceCube neutrino telescope	48

4.2.1		The InIce detector	48
4.2.2		The IceTop detector	50
4.2.3		The DeepCore extension	50

5 A stacking analysis with AMANDA-II and IC-22 data — 51
- 5.1 The source stacking method … 51
- 5.2 The source samples … 53
- 5.3 Analysis of AMANDA-II data … 55
  - 5.3.1 Optimization … 55
  - 5.3.2 Data … 58
  - 5.3.3 The optimized parameters … 58
- 5.4 Analysis of IC-22 data … 59
  - 5.4.1 Data … 59
  - 5.4.2 Changes compared to the AMANDA analysis … 60
  - 5.4.3 Changes of the source selection … 61
  - 5.4.4 The optimized parameters for IC-22 … 62
- 5.5 Results … 62

6 Conclusions & outlook — 67

Appendices — 69

A The source catalog of Starburst-Galaxies — 69
- A.1 General data … 69
- A.2 Radio data … 73
- A.3 Far Infrared data … 80
- A.4 X-ray data … 84

B Supplementary plots — 89
- B.1 Optimization plots of the AMANDA analysis … 89
- B.2 Optimization plots of the IC-22 analysis … 109

C Source lists — 119
- C.1 Sources lists for the AMANDA analysis … 120
- C.2 Sources lists for the IC-22 analysis … 125

List of figures — 129

List of tables — 133

Biblography — 135

1

Introduction

ince the beginning of mankind humans have explored the world in which they are living. Some questions were answered while new ones arose. One of the remaining questions is the composition of matter. In the ancient world philosophers tried to answer this question but it was not possible for them to experimentally confirm their findings. In the 16th century, the science of modern physics was started to develop, based on the idea to falsify or verify predictions experimentally and trying to find answers to how matter was built up. Throughout the centuries, knowledge was gained and different branches of physics developed to tackle the different arising questions. One relatively new branch is astroparticle physics acting as an interface between particle physics and astronomy. The field of astroparticle physics tries to answer the question about origin and composition of the high-energy cosmic radiation hitting Earth each second — the *cosmic rays*. Discovered over one hundred years ago the exact origin of cosmic rays is still unknown, since the charged particles are deflected by cosmic magnetic fields and do not reveal any directional information. Experiments in astroparticle physics evaluate the information carried by messenger particles. One of these messenger particles is the neutrino. Experiments which search for neutrinos from astrophysical sources are AMANDA and IceCube, both detectors at the geographical South Pole.
The present thesis deals with one particular source candidate, the Starburst-Galaxy. As part of this thesis a catalog of Starburst-Galaxies was collected and a source stacking analysis with data from AMANDA and IceCube was performed with the aim to detect neutrinos from Starburst-Galaxies and active galactic nuclei (AGN) or set constraints on a possible neutrino flux from these sources.
This thesis is organized as follows: The first chapter gives a brief overview of the status of astroparticle physics, the particle acceleration, possible sources of cosmic rays. The messenger particles and their detection are discussed. The second chapter introduces the Starburst-Galaxies and presents the source catalog. In chapter three the detectors used for the analyses done in this thesis are introduced. The next chapter presents the analyses of Starburst-Galaxies and AGN performed with AMANDA data and IceCube data, the results of these analyses are shown. The

last chapter contains conclusions and outlook. The appendices contain the source catalog of Starburst-Galaxies (appendix A), supplementary plots for the analyses (appendix B) and the source lists used in the analyses (appendix C).

2

Status of astroparticle physics

2.1 Brief historical development

The field of astroparticle physics is fastly developing having its roots in the discovery of cosmic rays as source of ionizing particles in 1912. Preceeding considerations by Elster, Geitel and Wilson [EG07, Wil01] about the conductivity of air led to the assumption around 1900 that there has to be an additional source for ionizing particles other than the radioactivity of the soil. An evidence to this alternate source was provided by Victor F. Hess in 1912. The aim of Hess' work was to research the change of radiation with altitude using balloon rides. The measurements were done with a charged electrometer which discharged within a time interval Δt due to the influence of radiation. Here, Δt depends on the intensity of the signal. If the radiation were originating from the Earth, a decrease in the ionization level would be expected. It was found by Hess that the intensity of the radiation decreased only to an altitude of a few hundred meters above the ground. With increasing altitude the intensity increased again to reach the same level as on the ground at an altitude of ~ 1800 m. The measurements were done up to an altitude of 5000 m, a higher radiation intensity than on the ground was detected. The results of Victor Hess lead to the conclusion that next to the known sources of ionizing radiation there must be other sources which have to be searched beyond the Earth [Hes11, Hes12]. Later this ionizing radiation was labeled *cosmic rays*. In the following period many particles were discovered in the cosmic rays, the positron for example [And33].

With the availability of powerful particle accelerators the interest in cosmic rays as an object to research faded. With these particle accelerators the particles could be studied under laboratory conditions. However, during the past decade the interest in cosmic rays increased again since no man made accelerator can accelerate particles to energies as cosmic accelerators can. For a review of the history of astroparticle physics see [Cir08].

2.2 The energy spectrum of cosmic rays

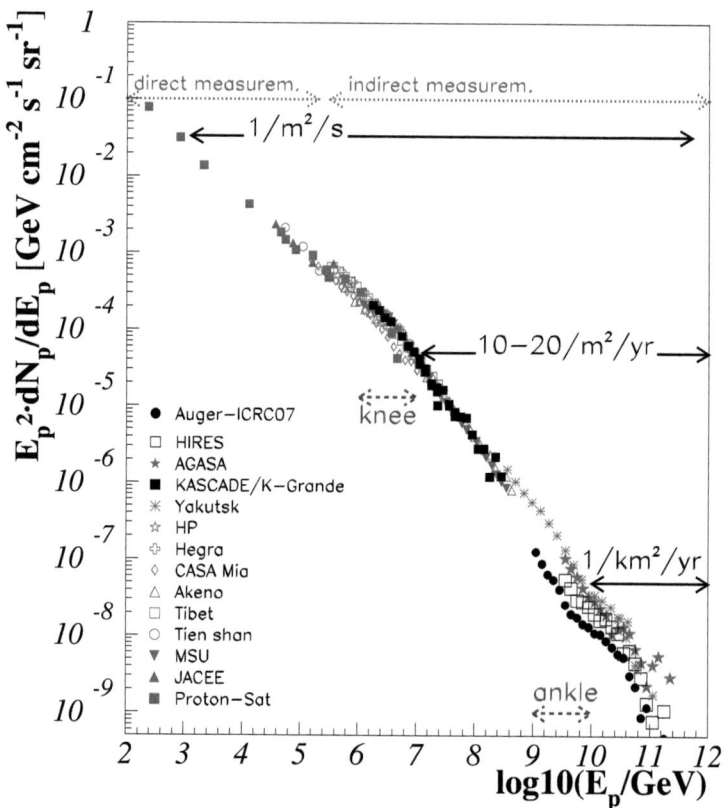

Figure 2.1: *The energy spectrum of cosmic rays measured by various experiments. The flux has been weighted with E^{-2}. The numbers on the right hand side is the expected number of particles on Earth per unit time [Bec08].*

Figure 2.1 is a presentation of the energy spectrum of cosmic rays measured by various experiments. It spans over ten orders of magnitude in energy i.e. from keV to EeV. The spectrum follows a power law, the particle flux is proportional to E^γ where γ is called spectral index which is a constant over large ranges of energies. However, there are two kinks in the spectrum: The first at $E_{\text{knee}} \approx 2.5 \cdot 10^{15}$ eV, here γ changes from 2.63 to 3.03 [Wie98]. This kink is called the 'knee' of the cosmic ray spectrum. The second kink is at an energy of $\sim 10^{18.5}$ eV, the spectral index changes to ~ 2.75 [Sta04]. This second kink is called 'ankle'. Above an energy of $\sim 10^{20}$ eV it was expected that the charged cosmic rays interact with the cosmic microwave background (CMB). Thus only few reach the Earth. In an interaction at such an energy of a proton with a photon a delta resonance is produced which then decays into a neutron (or proton) and a charged or neutral pion:

$$p + \gamma \to \Delta^+ \to \begin{matrix} n + \pi^+ \\ p + \pi^0 \end{matrix}.$$

The energy of the produced proton is always lower than the energy of the primary proton. The interaction of high energy cosmic rays with the CMB was predicted simultanously by the American physicist K. I. Greisen and Russian physicists G. T. Zatsepin and V. A. Kuzmin [Gre66, ZK66]. The expected cutoff in the cosmic ray spectrum due to the interactions with the CMB is therefore called *Greisen-Zatsepin-Kuzmin Cutoff* or short *GZK-Cutoff*. After the prediction in the year 1966 it was claimed to be experimentally confirmed by the Auger experiment [A+08a] and the HiRes experiment [B+08] 40 years later.

2.3 Particle acceleration

Cosmic rays are accelerated in astrophysical environments which are often characterized by the collision of plasmas. In a supernova explosion the shell of matter is blown from the central object and encounters the interstellar medium. Here, a shock front is produced since the gas in the shell encounters other gas with a velocity faster than any signal velocity. The supernova remnant (SNR) can be observed for more than 1000 years. This phenomena is not only limited to astrophysical environments. As an example, shocks are formed in supersonic movements of planes or bullets in air. In these examples the plane or bullet is moving faster than the speed of sound, which is the characteristic speed of the medium, and produces a shock wave, the Mach cone [Mac97, MW84, MW85]. In astrophysical environments the characteristic speed of plasma is the speed of magnetic waves.

Non-thermal emission of electromagnetic radiation at radio to X-ray energies can be interpreted as synchrotron radiation from shock-accelerated electrons. At energies above X-rays, the signal can arise from inverse Compton scattering of the highly-relativistic electrons with the ambient low-energy photon field. If a hadronic component is accelerated, the electromagnetic emission in the MeV to TeV range can also arise from protons. These protons can lose energy in synchrotron radiation or by proton-photon and proton-proton interactions. The latter two processes lead to photon emission from π^0 decays.

Stochastic particle acceleration as an acceleration mechanism that produces the observed power law behavior was first discussed by Fermi [Fer49, Fer54] and Darwin [Dar49]. In this mechanism a charged particle enters an acceleration region with an energy E_0 and leaves the acceleration region with an energy $E > E_0$.

There are two basic configurations for this accelerating mechanism. One is to consider a moving, partially ionized gas cloud as shown in figure 2.2(a). The incident particle has the energy E_1, the momentum p_1 and the angle θ_1, the according quantities for the outgoing particle are E_2, p_2 and θ_2. Magnetized interstellar gas clouds have typically a speed of $\sim 15\, \frac{\text{km}}{\text{s}}$ [Pro98]. The particles are scattered off the magnetic field inhomogeneities inside the cloud and leave the cloud with randomized directions. In this case the momentum gain per encounter with a magentic field inhomogeneity ξ is proportional to the squared cloud velocity, $\xi \propto V_c^2$, therefore this configuration of the mechanism is referred to as *second order Fermi acceleration*. Because the energy gain is of second order this process is not very efficient. In a different configuration a plane as infinite shock front is considered. This configuration is called *first order Fermi acceleration* and it is illustrated in figure 2.2(b). A particle with the inertial energy E_1 only changes its energy by crossing the shock. In this shock it is accelerated at moving magnetic field inhomogeneities to an energy E_2. The incident angle is labeled θ_1 in the figure, the angle of the returning particle towards the shock is labeled θ_2. In this configuration first order effects are dominant and the total momentum gain for one cycle of acceleration is $\xi \propto V_R$, V_R is the relative speed between the particle and the shock. A review of both acceleration mechanisms is given in [Pro98].

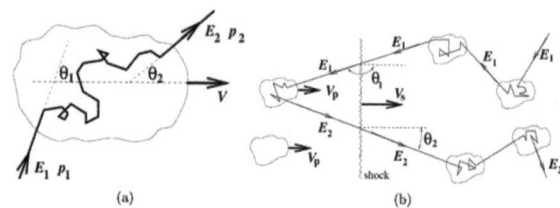

Figure 2.2: *On the left (a) there is an illustration of the second order Fermi mechanism, the more effective first order mechanism is displayed on the right side (b) [Pro98].*

2.4 Possible sources of cosmic rays

The acceleration mechanisms described in section 2.3 can occur in any astrophysical configuration where shock fronts and magnetic field inhomogeneities occur. This is the case in many astrophysical objects, though, not every object where Fermi acceleration occurs can accelerate particles to the observed energies. The ability of an object to accelerate charged particles to a

2.4. Possible sources of cosmic rays

certain energy depends on its size and magnetic field strength:

$$B \cdot L > \frac{2E}{Z \cdot e \cdot \beta}$$

This relation was first introduced by A. M. Hillas [Hil84], the magnetic field strength B is in units of μG, the size of the acceleration region L is in pc and the energy in units of 10^{15} eV. This relation is visualized in figure 2.3. On the x-axis there is the size L of the acceleration region, on the y-axis the magnetic field strength B. A source that is able to accelerate protons to 10^{15} eV ('knee' line) has to be left above this line in the figure, the same for energies above $10^{18.5}$ eV, ('ankle' line) or above 10^{20} eV, the GZK cutoff energy. Multiple source candidates are displayed in the figure which is referred to as the Hillas diagram or Hillas plot. From figure 2.3 it is visible that the cosmic rays below the ankle ($E < 3 \cdot 10^{18.5}$ eV) could be just from galactic origin.

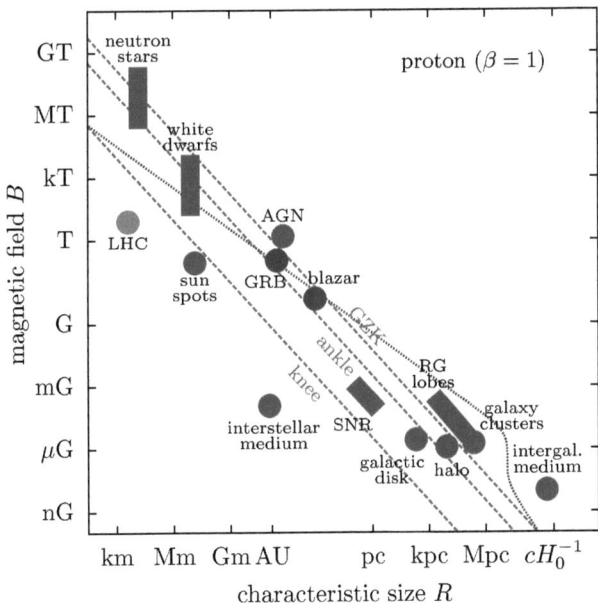

Figure 2.3: The Hillas diagram: The lines denote 10^{15} eV (knee), $10^{18.5}$ eV (ankle) and 10^{20} eV (GZK). A source that is able to accelerate particles above one of these energies has to be left and above the according line. Diagram taken from [A+09c] modified from [Pro04].

2.4.1 Supernovae and Supernova Remnants

It is commonly believed that the cosmic ray spectrum at energies below the knee is produced in expanding shells of supernova remnants (SNR). These SNR consist of shock fronts built by material that was ejected into the interstellar medium by a supernova explosion (SN). The supernova rate in the Galaxy is one SN per 50 years [CT01]. However, there are galaxies which have an increased supernova rate due to their increased star formation rate. These galaxies are called *Starburst-Galaxies*, they will be discussed in more detail in chapter 3. Although SN are good candidates for the production of cosmic rays it is difficult to explain the break in the spectrum at $E \sim 10^{15}$ eV. One possible explanation would be the leakage of particles out of the Galaxy leaving only heavy elements at higher energies. This may lead to a steepening of the spectrum. Another possibility is that SN exploding into their own wind are able to accelerate particles to higher energies since heavier elements (helium up to iron) are produced. A SN losing the hydrogen envelope before collapsing leads to a higher density of particles when the shock forms and thus to different conditions in the shock. Regular SNRs are expected to produce cosmic rays up to the knee while SNR winds are believed to accelerate particles up to the ankle [SBG93].

2.4.2 Pulsars and binary systems

There are also alternative explanations for the spectrum above the knee, systems with neutron stars or black holes are considered. Neutron stars are observed as pulsars, their electromagnetic radiation follows the magnetic field lines which are not aligned with the rotational axis, the emission can only be seen when the particle jet points towards Earth. The periodic signal from pulsars can range from several seconds to milliseconds. The most prominent since the most luminous pulsar is the Crab pulsar, a millisecond pulsar. The Crab pulsar is a neutron star which was produced in a SN explosion observed on 4th of July 1054 [Duy42, MO42]. Newer evaluation of historical records of this SN suggest an earlier date, 11th of April 1054 [CCM99]. The remnant of the SN 1054, the Crab nebula, has been observed in all wavelengths. It was seen from radio up to TeV energies, the pulsar itself is most luminous at X-ray and higher energies. Pulsars have very high magnetic fields of $B \sim 10^{15}$ G which makes them good candidates for particle acceleration.

Another good candidate for shock acceleration is the class of binary systems including a neutron star or a black hole. Low-Mass X-ray binaries (LMXBs) consist of a white dwarf and a companion star while High-Mass X-ray binaries (HMXBs) consist of neutron star that is fed by a blue (O/B) star. If in a HMBX the companion exceeds the Roche volume of the binary system it begins to feed the compact object with matter.

2.4. Possible sources of cosmic rays

The neutron star or black hole in turn emits the gained energy in a jet along the magnetic axes. These system can lead particle acceleration up to the ankle at most. See [Gai90] for a summary of X-Ray binaries and cosmic rays.

2.4.3 Active galactic nuclei

The cosmic rays at energies above the ankle cannot be of galactic origin since they are too isotropic. A promising source candidate in this energy regime are active galactic nuclei (AGN). This class of galaxies has a particularly bright core. The first AGN was discovered in 1962 as an object with a star like core. Since it showed extreme radio emission features it could not be classified as a star. The suggestion that this object which is today known as 3C 273 was indeed a distant galaxy with a bright core was first suggested by Maarten Schmidt one year after the discovery [Sch63]. These objects used to be referred as quasi stellar objects (QSOs). Today it is known that QSOs fit into the general classification scheme for AGN. AGN are objects which are believed to be powered by a rotating super massive black hole in the center. A schematic

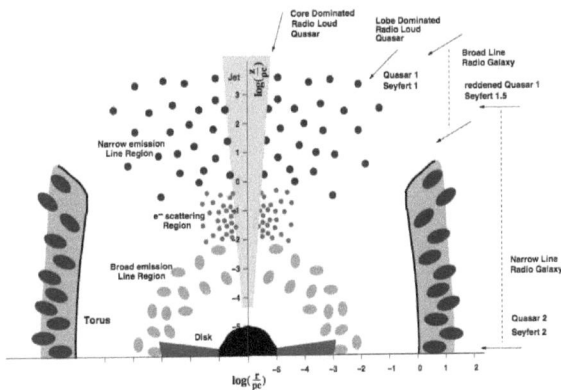

Figure 2.4: *Schematic view of a cylindrical symmetric AGN in the r-z-plane. Both axis are logarithmically scaled to 1 pc. It is indicated which objects are believed to bee seen from which direction. Figure from [ZB02].*

view in an AGN is shown in figure 2.4. AGN are called "active" due to the accretion disk which forms around the central black hole and radiates strongly in optical bandwidths. This accretion disk is fed with material from a dust torus. Perpendicular to this disk there are two relativistic jets emitted, transporting matter in the form of lobes. In these lobes there are knots and hot spots which emit radio emission, leading to the strong radio signal observed in AGN. It is believed that these knots and hot spot represent shock acceleration regions in which particles are accelerated to high energies, protons for example up to $E_p \sim 10^{21}$ eV [BS87]. Recent results from the Auger experiment suggest a correlation between high energetic cosmic rays and AGN. A correlation between cosmic rays with energies above $6 \cdot 10^{19}$ eV and the position of AGN closer than 75 Mpc was demonstrated [AP+07, AP+08]. The observed types of AGN fit into the AGN classification scheme (figure 2.5). In this scheme three main criteria were used to classify the AGN:

1. The AGN are divided into radio loud and radio weak AGN according to their activity in radio wavelengths. About 90% of all AGN are radio weak and are usually hosted in spiral galaxies while the radio loud AGN are hosted in the centers of elliptical galaxies.

2. A further classification criterion is the optical luminosity. The radio weak sources are subdivided into optically strong and optically weak, this can be distinguished by considering the features of the emission lines. Narrow emission lines are usually missing in optically strong sources while they are present in optically weak sources. Broad emission lines are present in both source types. Radio loud sources with extended jets ($\sim 100\,\mathrm{kpc}$) are further subdivided at radio wavelengths into low and high luminosity. The critical luminosity is $L_\nu = 2.5 \cdot 10^{26} \frac{\mathrm{W}}{\mathrm{Hz}}$. The jets of compact sources like GHz-Peaked-Sources (GPS) and Compact Steep Sources (CSS) are believed to get stuck in matter.

3. The last criterion is the orientation of the AGN towards the observer. AGN are axisymmetric along the axis of the jet. An object is classified as blazar in the branch of radio loud AGN if one of the jets points directly to the observer. Blazars are further divided into Flat Spectrum Radio Quasars (FSRQs) if they have high luminosity and into BL Lacs if they have low luminosity. The Faranoff Riley (FR) galaxies are being looked at from the side, and thus the torus and the jet are clearly visible. These galaxies are divided according to their luminosity, too. The high luminosity population is labeled FR-II galaxies showing very high radio emission at the end of the jets, while the radio emission of the low luminosity population class happens in knots throughout the jet. This class is labeled FR-I.

The radio weak AGN are called radio weak quasars in the optically strong case. The optically weak objects are called Seyfert-I galaxies, when looked at the gap between jet and AGN torus. The equivalent to the FR galaxies in the radio weak case are Radio Intermediate Quasars (RIQ) and Seyfert-II galaxies, the observer's view is here directed towards the torus.

2.4. Possible sources of cosmic rays

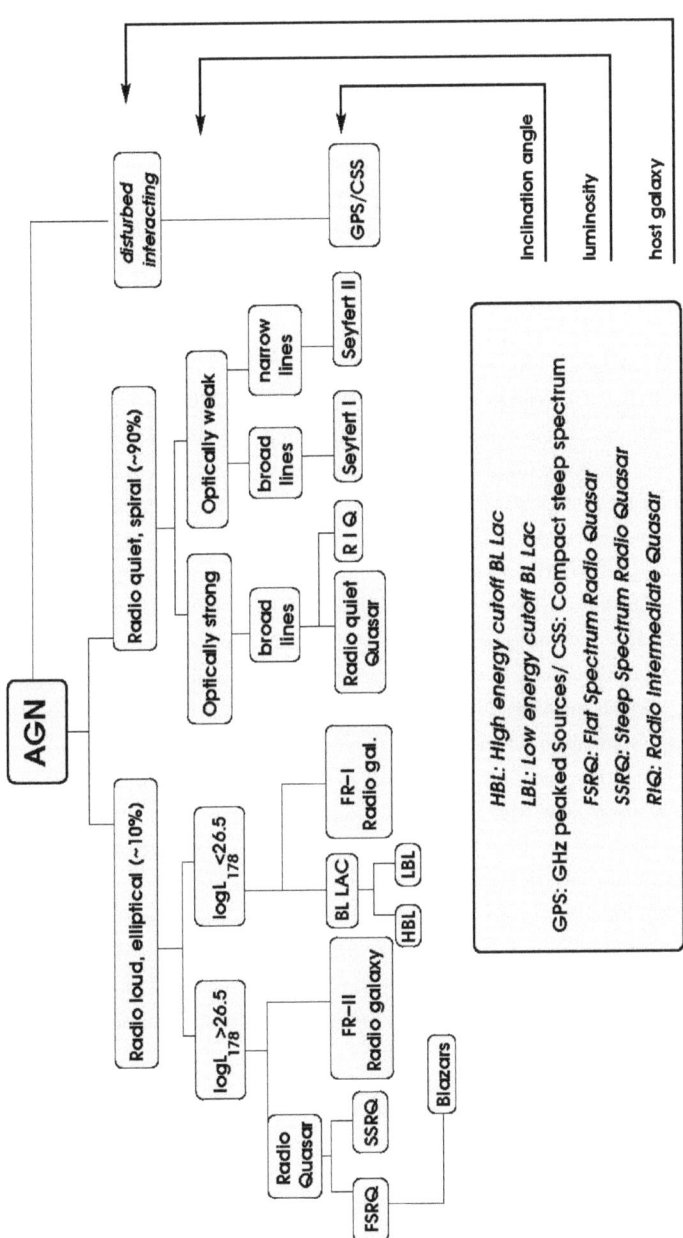

Figure 2.5: *The AGN classification scheme. Figure from [Bec08, A+06].*

2.4.4 Gamma ray bursts

In the 1960s photon eruption of unknown origin were detected both by American and Soviet military satellites. Although it was clear immediately that these events were not man made, it took until 1973 until this was first published after the first observation in 1967. This first publication was from the American Vela satellites [SKO73], a few months later from the Soviet Cosmos-461 [MGI74] and the American OSO-7[1] and IMP-6[2] satellites [WUB+73, CD73]. Further systematic studies of these Gamma Ray Bursts (GRB) were done with BATSE[3] on board the CGRO[4], BATSE was taking data for 9 years, from April 1991 until June 2000 [PMP+99]. BATSE detected 2704 GRBs in an energy range between 20 keV and 2000 keV. The bursts distribute isotropically with no visible clustering in the galactic plane or anywhere else suggesting that GRBs are not of galactic origin. While the prompt emission of the GRBs is in the keV – MeV range, the so-called afterglow emission continues long after the prompt emission and is seen in practically all wavelengths. From the observation of the afterglow host galaxies can be identified and the distance of the GRB through its redshift can be determined [WRM97]. This was first done in 1997, when the satellite BeppoSAX[5] detected GRB970228 and also measured the GRB afterglow in X-rays. This detection and follow up observations in optical and other wavelengths made the determination of the redshift possible. The redhift of this GRB was determined to $z = 0.9578$ which corresponds to ~ 6 Gpc [D+01]. This was the proof that GRBs are not of galactic origin. GRBs have a bimodal duration distribution, the duration of a GRB is labeled t_{90}. Here, t_{90} is the time span in which 90% of the signal was received. The duration distribution has two populations, one with a $t_{90} > 2$ s classified "long" and one with $t_{90} < 2$ s classified "short". It has been unclear for a long time what is the mechanism that causes long and short GRBs. In 2003 it was discovered that long GRBs are connected to supernova explosions of type Ic [MDT+03]. Supernovae of type Ic follow the death of very massive Wolf-Rayet stars. Short GRBs have been proven in 2005 to originate from the merging of two neutron stars or a neutron star and a black hole in a binary system [HSG+05, VLR+05, GCG+06]. Long and short GRBs do not only differ in their duration but also in their redshift distribution. While long GRBs happen in star forming regions following the star forming rate, short bursts happen in regions with rather low star formation rate and at small redshifts ($z \sim 0.1$) [Bec08].

---

[1] **O**rbiting **S**olar **O**bservatory
[2] **I**nterplanetary **M**onitoring **P**latform
[3] **B**urst **A**nd **T**ransient **S**ource **E**xperiment
[4] **C**ompton **G**amma **R**ay **O**bservatory
[5] Name made out of the name of the Italien astronomer Giuseppe Occhialini called **Beppo** and '**SA**tellite per lo studio a raggi **X**', Italian for 'Satellite for X-ray studies'

2.5 Messenger particles and their detection

From the possible sources of cosmic rays different particles reach the Earth and make the observation of the sources possible through different techniques of detection. The cosmic rays at an energy above 1 GeV consists of ~ 98% nucleons and ~ 2% leptons [Lon92]. The particles and the detection technique used for detection are discussed in the following and are shown graphically in figure 2.6.

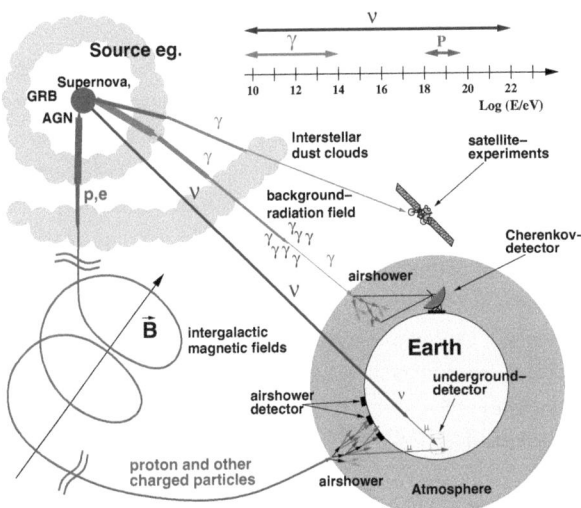

Figure 2.6: A graphical overview of propagation and detection of the different particles of the cosmic rays [Wag04].

2.5.1 Protons and heavy nuclei

Protons and heavy nuclei are often referred to as primary cosmic rays since they are accelerated in the source regions and reach the Earth without interaction. Above 1 GeV the nucleons in the cosmic rays are ~ 87% protons, ~ 12% α-particles and ~ 1% heavy nuclei [Lon92]. Although the primary cosmic rays carry valuable information from the source, like the energy of the particles and the energy spectrum which could reveal the properties of the source and the acceleration mechanism, they lack an also very important information: the direction. Charged cosmic rays are deflected in interstellar magnetic fields on their way to Earth and thus do not contain any information anymore where they came from and thus can not be correlated to the source objects. However, this might not be true for charged particles at the highest energies. If particles originate from nearby sources and have sufficient energies, they can travel on almost straight lines and could be correlated to source objects. This was already mentioned for active galactic nuclei, see subsection 2.4.3. The exact values for the maximum distance and the minimum energy depend on the magnetic field configuration, which is still not well-known.

Detection of protons and heavy nuclei

The direct measurement of the primary cosmic rays is only possible outside the Earth's atmosphere since the charged particles interact in the atmosphere. Thus detectors for direct measurements are on board satellites or balloons. The first dedicated satellite borne experiment for the detection of cosmic rays is PAMELA[6] on board the Russian Resurs-DK1 satellite. It is designed to measure different kinds of matter and antimatter in the cosmic rays such as protons, electrons and their anti particles [C+08]. A similar balloon based experiment was ATIC[7]. ATIC flew several flights in Antarctica and measured protons and nuclei [I+09]. Since space is limited on satellites and balloons these experiments have only small active detector volumes which limits them to small energies, $< 1\,\text{TeV}$ for protons in the case of PAMELA. If charged particles interact in the atmosphere they produce cascades of secondary particles. These cascades can contain just leptons in case of an electromagnetic cascade or they may contain mesons in case of a hadronic cascade. Next to the secondary particles also fluorescence and Čerenkov light is produced in an air shower, both can be used for indirect detection of primary cosmic rays. In figure 2.7 a schematic view of an air shower is shown. There are several techniques to detect air showers:

---

[6]**P**ayload for **A**ntimatter **M**atter **E**xploration and **L**ight-nuclei **A**strophysics
[7]**A**dvanced **T**hin **I**onization **C**alorimeter

2.5. Messenger particles and their detection

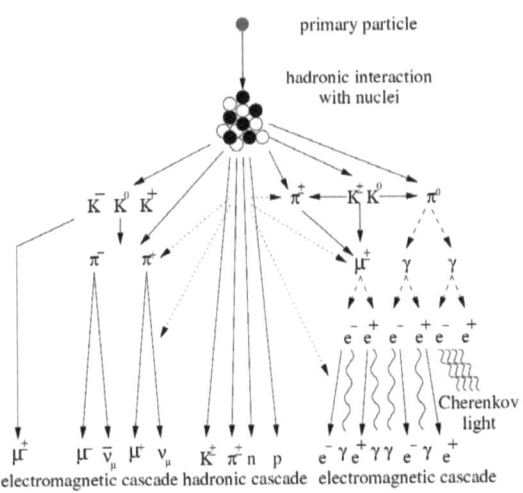

Figure 2.7: *Schematic view of an air shower produced by a primary particle like a proton or a heavy nucleus [Wag04].*

- **Ground detectors** detect secondary particles induced from primary particles interacting with the atmosphere. From the lateral and time distribution of these particles the shower the direction of the primary particle is estimated. From the amount of secondary particles detected the energy of the primary particle is estimated. The ground detectors are scintillation counters, muon counters or photomultipliers (PMT) ordered in an array. An example for an array with scintillation counters is the KASCADE-Grande[8] experiment which stopped data taking in March 2009. It consisted of 237 scintillation detectors and was designed to measure air showers with energies between 500 TeV and 1 EeV [S+09].

- **Fluorescence light** is caused by scattering processes of secondary electrons and positrons in the atmosphere. The primary source for fluorescence light in the atmosphere is the excitation of nitrogen molecules. One experiment that used the fluorescence technique is HiRes[9] in Utah which operated from June 1997 until April 2006. The experiment consisted of two fluorescence telescopes which allowed stereoscopic observations and thus better reconstruction of the shower geometry. HiRes has confirmed the GZK cutoff [B+08]. Currently as a successor for HiRes the Telescope-Array at the same site has just begun datataking [N+09]. There are also experiments which use both techniques, ground detectors and fluorescence detectors. The Pierre Auger Observatory in Argentina uses 1660 water Čerenkov detectors and 24 fluorescence telescopes in groups of 6 telescopes at 4

[8] **KA**rlsruhe **S**hower **C**ore and **A**rray **DE**tector-Grande
[9] **Hi**gh **Res**olution Fly's Eye

locations at the edge of the surface detector array [A+09e]. Like HiRes the Pierre Auger Observatory has also confirmed the GZK-cutoff [A+08a].

- **Čerenkov light** is emitted when a charged particle passes through a medium with a velocity faster than the speed of light in that media. The Čerenkov light is emitted in a cone with an opening angle θ_C proportional to the speed of the particle. With $n = \frac{c_0}{c}$ and $\beta = \frac{v}{c_0}$ θ_C is given by

$$\theta_C = \frac{1}{n \cdot \beta} . \qquad (2.1)$$

Here, c_0 being the speed of light in vacuum. However, equation 2.1 is in fact an approximation to

$$\cos \theta_c = \frac{1}{\beta \cdot n} + \frac{\hbar k}{2p}\left(1 - \frac{1}{n^2}\right). \qquad (2.2)$$

Here, p is the momentum of the particle and $\hbar k$ the momentum of the emitted photons. Since $\hbar k \ll p$ equation 2.1 is a good approximation for equation 2.2 [Gru93]. In figure 2.8 there is a sketch of the geometry of the Čerenkov effect. The emission of Čerenkov light happens because the particle produces dipoles on its path through the medium. The dipole fields interfere constructively if the speed of the particle is larger than the phase speed of light in the medium [Jac62]. If the speed is less than this, the interference is destructive and no Čerenkov light is emitted.

2.5. Messenger particles and their detection

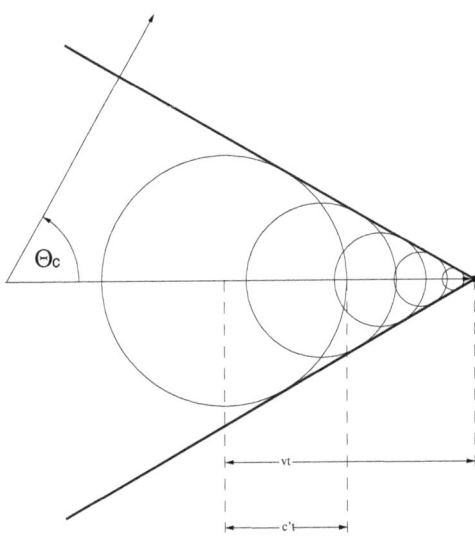

Figure 2.8: *The Čerenkov effect: If the velocity of a prticle is faster than speed of light in the medium the induced dipoles interfere cintrutctively and Čerenkov light is emitted in a cone with an opening angle θ_C. Figure from [Jac62].*

The Čerenkov light can be observed in air showers of primary particles and it can also be used to detect neutrinos. Neutrino detection will be discussed later in this thesis. The Čerenkov light of air showers is observed with imaging air Čerenkov telescopes (IACTs). The light produced in an air shower is observable for nanoseconds. Thus IACTs use a mirror to collect the light and PMTs to detect it in a PMT camera. There are numerous IACTs, for instance three of them are mentioned here. The MAGIC-Telescope[10] is situated on the Canary island La Palma. MAGIC has two individual telescopes, MAGIC-I and MACIG-II, each with a mirror diameter of 17 m, being the world's largest single dish IACTs. MAGIC-II is operable since April 2009. The MAGIC telescopes are designed to detect very high energy (VHE) γ-rays in an energy range between 25 GeV and 20 TeV [A+08b, A+09h]. Similar to MACIC there are the H.E.S.S.[11] telescopes in Namibia and VERITAS[12] in Arizona. H.E.S.S. is an array of four telescopes with a diameter of 13 m each (H.E.S.S. Phase I) and is currently extended by another larger telescope (25 m diameter) in the middle of the array (H.E.S.S. Phase II). The lower energy threshold of Phase I is 100 GeV, for Phase II this is expected to be lowered to $\sim$ 20 GeV [Hor07]. VERITAS is an array of four telescopes, each telescope having a mirror diameter of

---
[10] **M**ajor **A**tmospheric **G**amma-ray **I**maging **C**herenkov **Telescope**
[11] **H**igh **E**nergy **S**tereoscopic **S**ystem
[12] **V**ery **E**nergetic **R**adiation **I**maging **T**elescope **A**rray **S**ystem

11 m [The09, MHKM09]. VERITAS has a detection threshold of 100 GeV for photons and it is planned to upgrade the system to lower this threshold by relocating one telescope and improving the camera and trigger system [The09, Ott09]. Although the mentioned IACTs are optimized for the detection of VHE photons it is also possible to detect protons which form the background in VHE photon observations.

- **Radio detection** of cosmic rays is known for about 40 years but has been almost forgotten in the early 1970s since technical difficulties were encountered and other detection techniques seemed to be more promising at that time. The radio detection technique tries to detect the synchrotron radiation of electron/positron pairs in air showers caused by the Earth's magnetic field. This *geosynchrotron radiation* is a strongly pulsed radio emission at frequencies around 40 Hz. It can be detected in frequency ranges between a few and a few hundred MHz. Today, radio detectors are added to existing detector arrays like the LOPES[13] array of dipole antennas as an extension of the KASCADE-Grande array. LOPES is a prototype for a larger array planned to be spread over the Netherlands and parts of Europe which is called LOFAR[14]. For an overview of geosynchrotron radiation and radio emission of extended air showers see [HF03].

2.5.2 Photons

Photons from cosmic ray sources are produced in the sources by interaction processes of the accelerated particles. Photons cover a vast range in energy, from radio radiation with a few eV to VHE photons with several TeV or even EeV. While photons are not deflected by magnetic fields, they are more or less easily absorbed by interstellar clouds or matter within the source itself. Thus photons may only carry information about the surface of the source. The observation of photons can be done directly or indirectly. Indirect measurements were already covered in the subsection before. High energy γs can be observed directly with satellite or balloon experiments and indirectly by observing the air showers produced by them in the atmosphere. Only a small part of the photon spectrum can be observed from the surface of the Earth since the Earth's atmosphere is opaque for all frequencies except the visible light (optical window) and radio waves (radio window).

Visible light observations

The oldest branch of astronomical observations is the observation of the visible light, it is as old as mankind. The first observations were done with the naked eye then later with optical telescopes[15]. The first optical telescope is believed to be built by the Dutch Hans Lipperschey

---
[13]**LOFAR PrototypE Station**
[14]**LOw Frequency ARray**
[15]derived from the Greek word $\tau\eta\lambda\epsilon\sigma\kappa\acute{o}\pi o\varsigma$ meaning 'to look far'

2.5. Messenger particles and their detection

in 1608 and one year later by Galileo Galilei [Kin55]. Today optical telescopes use mirrors with diameters up to 10.4 m in the case of the Grand Telescopio Canarias (GranTeCan), like the MAGIC telescopes on the island of La Palma. The GranTeCan had its first light in 2007 and is currently being equipped with different instruments used for spectroscopy in different wavelengths [RA08].

Radio observations

Another also quite old branch in the astronomy is the radio astronomy. The first radio signals from outside the solar system were detected in 1933 by Karl Guthe Jansky[16] with a turnable antenna. It turned out, that the radio signals he recorded were coming from the Galaxy with a maximum at the galactic center [Jan33]. The first sky survey in radio wavelengths was done by Grote Reber and published 1944 showing a map of radio emission from the Galaxy [Reb44]. In the 1950s it was found out that the radio emission was non thermal synchrotron radiation [Bur56], stemming from accelerated charged particles in the magnetic fields of the sources. Radio telescopes usually have a dish shaped reflector that focuses the radio waves onto a receiver. Usually telescopes have multiple receivers for observations at different frequencies. Radio telescopes have rather large reflector dishes, the diameter ranges from a few meters to several hundrets of meters. For example the free steerable radio telescope in Effelsberg has a diameter of 100 m. The largest radio telescope is the RATAN 600[17] telescope located in the north Caucasus, Russia. The 600 stands for the diameter of the reflector which is about 600 m. This reflector is not a dish but a ring of reflecting panels which reflect the signals to a central cone shaped reflector which focuses the signal. Due to this design RATAN 600 has only limited possibilities to select the telescopes field of view. To improve the sensitivity and the angular resolution if radio telescopes interferometry of multiple telescopes is used. For example the VLA[18] in New Mexico which consists of 27 dish antennas with a diameter of 25 m arranged Y-shaped on rails to obtain a flexible array. Interferometry with radio telescopes is also done on large scales with baselines of thousands of kilometers. As an example to mention here is the VLBA[19]. The VLBA uses 10 antennas spread over the US. The longest baseline between two antennas is 8116 km.

Infrared observations

The observations in other wavelengths than radio, visible light and the near infrared are mostly done with satellites, since other wavelengths are absorbed by the atmosphere. The first instrument for dedicated observation of the far infrared was IRAS[20] which did a sky survey with 98%

---
[16] today radio astronomers measure the radio flux density in units of Jansky: $1\,\text{Jy} = 10^{-26}\,\frac{\text{W}}{\text{m}^2\text{Hz}}$
[17] cyrillic PATAH 600 - Radio Telescope of the academy of sciences
[18] Very Large Array
[19] Very Long Baseline Array
[20] InfraRed Astronomical Satellite

coverage of the whole sky and operated from January to November 1983 [WGC+94]. IRAS detected hundreds of thousands of sources in four wavelengths, 12 µm, 25 µm, 60 µm and 100 µm. Infrared satellites have duty cycles which are limited by the supply of coolant, liquid hydrogen or liquid helium, on board. The infrared detectors have to be cooled down to a few K to be able to detect small fluxes. The infrared satellite currently operating is the Spitzer Space Telescope or short Spitzer[21]. Spitzer observes wavelengths between 3 µm and 180 µm and was launched in 2003. It uses a heliocentric orbit since in such an orbit the sun is always in the same direction from the satellite and makes effective heat shielding possible [AR06]. After observing for over five-and-a-half years the liquid helium supply was exhausted, since then Spitzer is operating at a temperature of about 30 K leaving the spectrometers for the short wavelengths operable [SSWS09].

Ultraviolet observations

One of the first instruments that was designed for observations of astronomical objects in the ultraviolet (UV) light was on board the European TD-1 satellite which was launched 1972. TD-1 was also capable of measuring charged particles and X-rays. TD-1 did a UV sky survey [B+73, T+78]. Which was the only UV sky survey until 2003 when GALEX[22] was launched. GALEX measures in two wavelength bands, 1350 − 1370 Å and 1750 − 2750 Å, and its primary mission is an imaging all sky survey [M+05]. The survey of TD-1 only covered stars.

X-ray observations

Cosmic X-rays are like UV and infrared radiations only detectable outside Earth's atmosphere. The first observation of X-rays coming from the sun were done 1949 with detectors attached to the nose of a V2 rocket. Later X-ray detectors were mounted to satellite experiments. The today most capable instruments are the Chandra X-ray observatory[23] and XMM-Newton[24]. Chandra was launched 1999 and is able to detect X-rays with energies between 0.08 keV and 10 keV. For focusing the X-rays Chandra uses hyperbolic glass mirrors coated with iridium. For a detailed description of Chandra see [WBC+02]. XMM-Newton was launched in 1999 and is operating now for ten years detecting X-rays with energies of 0.1 keV − 10 keV. Since XMM-Newton has three independent X-ray telescopes on board high resolution oservations and spectroscopy can be done at the same time [Güd09]. Next to dedicated X-ray telescopes there are multiple satellites which are also able of detecting X-rays they will be addressed next.

---

[21] named after Lyman Spitzer Jr., American astrophysicist, driving force for the development of space telescopes
[22] **GAL**axy **E**volution **EX**plorer
[23] named after the astronomer Subrahmanyan Chandrasekhar
[24] **X**-ray **M**ulti-**M**irror Mission - **Newton**, named after Sir Isaac Newton.

γ-ray observations

A lot of satellite experiments launched for astronomical observations were capable of detecting γ-rays with energies of MeV to ~ 100 GeV. Energies larger than 100 GeV are observed indirectly with IACTs, which was described in subsection 2.5.1. Because of the large variety of instruments only the latest instrument as an example will be mentioned here. The at the moment most promising instrument is the Fermi Gamma Ray Space Telescope, formerly known as GLAST[25] which was launched in June 2008. The main instrument on board the Fermi spacecraft is the Large Area Telescope (LAT) which is currently performing an all-sky gamma-ray survey from 30 MeV to 300 GeV. The LAT started operations in August 2008 [Ran09]. The second instrument is the Gamma-ray Burst Monitor (GBM). The GBM is designed to detect GRBs, extend the energy range of the LAT down to ~ 8 keV– ~ 40 MeV and to compute GRB positions to relocate the spacecraft to allow detailed studies of GRBs at high energies with the LAT [M+09].

2.5.3 Neutrinos from astrophysical sources

Yet another branch of astronomy is the neutrino astronomy. The neutrino was first postulated in 1930 in a private letter exchange between Wolfgang Pauli and the participants of a congress in Tübigen. Pauli derived the existence of the neutrino from the fact that the electron spectrum of the β decay was continuous. Pauli called the missing particle at that time 'neutron' it was then later renamed to 'neutrino' by Enrico Fermi to avoid confusion. The neutrino was experimentally verified by F. Reines and C. Cowen in 1956 using the inverse β decay [RC+56]

$$\bar{\nu}_e + p \to n + e^+ \ .$$

Reines and Cowen determined the anti neutrino interaction cross-section to $(11 \pm 2.6) \cdot 10^{-44}$ cm^2 per $\bar{\nu}$. Later investigations of the cross section revealed a nearly linear dependence of the cross section from the energy:

$$\sigma(\nu N) = (0.682 \pm 0.012) \cdot 10^{-38} \, \text{cm}^2 \cdot E_\nu \, \text{GeV}^{-1}$$

for neutrino-nucleon scattering and

$$\sigma(\bar{\nu} N) = (0.338 \pm 0.007) \cdot 10^{-38} \, \text{cm}^2 \cdot E_\nu \, \text{GeV}^{-1}$$

for anti-neutrino-nucleon scattering with $E_\nu < 10$ TeV [Sch97].

Neutrino production in astrophysical environments

High-energy neutrinos observed at Earth are produced in interactions of charged cosmic rays. The neutrinos themselves are not accelerated since they are neutral. The most important

---
[25] Gamma-ray Large Area Space Telescope

process for neutrino production is the decay of charged π mesons

$$\pi^+ \to \mu^+ \nu_\mu \to e^+ \nu_e \nu_\mu \bar{\nu}_\mu$$

as well as

$$\pi^- \to \mu^- \bar{\nu}_\mu \to e^- \bar{\nu}_e \nu_\mu \bar{\nu}_\mu \,.$$

In the sources the charged mesons are produced in hadronic interactions of protons, neutrons and photons:

$$\begin{aligned} pp &\to X + p \to p n \pi^+ \\ np &\to \Delta^0 p \to p p \pi^- \\ p\gamma &\to \Delta^+ \to n \pi^+ \\ n\gamma &\to \Delta^0 \to p \pi^- \,. \end{aligned}$$

2.5.4 Neutrino detection

Due to their small interaction cross-section and since they are not charged neutrinos may travel far distances from the sources without being absorbed like photons or deflected like charged cosmic rays. Neutrinos may even carry information from inside the sources. However, the advantage for the propagation turns into a challange for the detection. Although neutrinos from a cosmic source will interact with matter only rarely, there are ways to detect neutrinos:

Detection with scintillators

This detection technique uses elastic scattering or charged current interactions to detect electrons or positrons produced in these interactions with scintillators. As an example, the Borexino detector, placed in the underground at the Laboratori Nazionali del Gran Sasso (LNGS) in Italy, uses a liquid scintillator to detect solar neutrinos. The detector uses elastic scattering and detection of the recoil electrons. Borexino uses 200 t of liquid scintillator as inner detector. The scintillator is a mixture of 1,2,4-trimethylbenzene(PC)[26] and 2,5-diphenyloxazole (PPO)[27]. Around this inner sphere there is another sphere which carries 2112 PMTs for detecting the scintillation light. The space between the inner and the outer sphere is filled with pure PC as a buffer liquid. Both spheres form the inner detector. The inner detector is placed in a water tank filled with ultra pure water and equipped with PMTs as a veto [Lew09]. Borexino has a threshold energy of 250 keV – 665 keV and can detect solar neutrinos having energies in the range of a few MeV. For higher energies much larger detector volumes are needed.

---

[26] $C_6H_3(CH_3)_3$
[27] $C_{15}H_{11}NO$

Radiochemical detection

Detectors with radiochemical methods to detect neutrinos use reactions where a target nucleus is transformed into a different nucleus in charged current interactions. Suited as target material is for example Gallium which is turned in the reaction

$$\nu_e + {}^{71}\text{Ga} \rightarrow {}^{71}\text{Ge} + e^-$$

into Germanium and an electron. A detector that uses Gallium as target is GNO[28]. GNO has a 100 t Gallium Chloride target which contains 30.3 t of Gallium. After an experimental run the Germanium is chemically extracted and the decays of the Germanium are counted. One decay corresponds to one neutrino interaction, ^{71}Ge decays with a half life time of 11.34 days. GNO has a detection threshold of 233.2 keV which makes it suitable for solar neutrinos [A+05].

Detection using the Čerenkov effect

Since the detection of neutrinos with higher energies requires large detector volumes, radiochemical detectors and scintillator detectors reach their limits in size. Larger detectors can be built using the Čerenkov effect (section 2.5.1) to detect neutrinos. In principle a Čerenkov detector uses a volume of interaction media which is transparent for light. This interaction can be a charged current or neutral current interaction. In case of a charged current interaction of a ν with a nucleon N a and lepton l is produced

$$\nu_l + N \rightarrow X + l \,.$$

Here, X indicates the hadronic product of the interaction which leads to a hadronic cascade. In case of a neutral current interaction a hadronic product and another neutrino is produced

$$\nu_l + N \rightarrow X + \nu_l \,.$$

In a detector the Čerenkov light of the leptons and the hadronic cascades can be observed using PMTs. As interaction and detection medium water or ice is suitable. A detector which uses ice is IceCube. IceCube will be discussed in detail in section 4.2.

Detectors using water can be either built as tanks filled with water or in open sea. An example for the first case is Super-Kamiokande[29], located in the Mozumi mine of the Kamioka Mining and Smelting Company near the village of Higashi-Mozumi, Gifu, Japan. The detector consists of a welded stainless-steel tank with 39 m diameter and 42 m tall with a total nominal water capacity of 50,000 tons. For the detection of the Čerenkov light it uses 11,146 PMTs facing to the inside and as a veto 1885 PMTs facing to the outside. The detector is capable of measuring neutrinos from 4.5 MeV to over 1 TeV [F+03]. The preceding experiment to Super-Kamikande,

---

[28] **G**allium **N**eutrino **O**bservatory
[29] Super-**Kamioka** **N**ucleon **D**ecay **E**xperiment

Kamiokande-II, did the first direct observation in neutrino astronomy in 1987. Kamioka-II detected neutrinos coming from the supernova SN1987A [HKK+88].

A proposed Čerenkov neutrino detector in the Mediterranean sea is KM3NeT[30] which is planned to have 1 km³ of detector volume instrumented. It is aimed that KM3NeT will be able to detect neutrinos with energies starting at a few hundred GeV to above 10 TeV with a pointing resolution of $0.1°$ [Rap08]. Currently there are three prototype detectors for KM3NeT, NEMO[31] [Dis09] NESTOR[32] [AAB+06] and ANTARES[33] [Bro09]. These projects are used to gain experience on the field of deep sea neutrino detectors, testing different configurations, hardware and sites for KM3NeT.

Acoustic and radio detection of neutrinos

To measure neutrinos at extremely high energies (EHEs) i.e. above 10^8 GeV, radio and acoustic detection techniques are being designed. The possibility that the Čerenkov effect was not just visible in visible light and UV but also in radio was first discussed by Askaryan [Ask62], therefore called *Askaryan effect*. It was successfully observed in sand [SGW+01], salt [GSF+05] and ice [GBB+07]. The Askaryan effect is used to detect EHE neutrinos at the South Pole with the AURA[34] experiment which is the radio extension to IceCube. The first radio receivers have been deployed to the ice and were successfully tested [LRV08].

The first experiment for acoustic neutrino detection was the SAUND[35] experiment which was deployed into the ocean [VGL05]. There are also acoustic extensions for ANTARES, the AMADEUS[36] experiment [Lah09], and for IceCube SPATS[37] [Des09]. The aim of these detectors is to test the feasibility of constructing an array of ~ 10 km³ for the detection of neutrinos in the EeV regime.

2.6 Astrophysical neutrinos

Two sources of astrophysical neutrinos are confirmed until today. One of these two sources is the sun. The observation of the neutrinos from the sun, which are produced in nuclear fusion processes in the sun, led to the discovery of neutrino oscillations, see e.g. [SNO02]. The second confirmed astrophysical neutrino source was the supernova SN1987A. From this supernova 24 events were observed within 13 s by five different experiments [Hel87]. In figure 2.9 there is an overview of the neutrino energy spectrum expected over a wide range of energy

---

[30] **k**m³ **N**eutrino **T**elescope
[31] **NE**utrino **M**editerranean **O**bservatory
[32] **N**eutrino **E**xtended **S**ubmarine **T**elescope with **O**ceanographic **R**esearch Project
[33] **A**stronomy with a **N**eutrino **T**elescope and **A**byss environmental **RES**earch project
[34] **A**skaryan **U**nder-ice **R**adio **A**rray
[35] **S**tudy of **A**coustic **U**ltra-high **N**eutrino **D**etection
[36] **A**NTARES **M**odules for **A**coustic **DE**tection **U**nder **S**ea
[37] **S**outh **P**ole **A**coustic **T**est **S**etup

2.6. Astrophysical neutrinos

from MeV up to EeV. The Cosmic Neutrino Background (CνB) is the equivalent to the cosmic

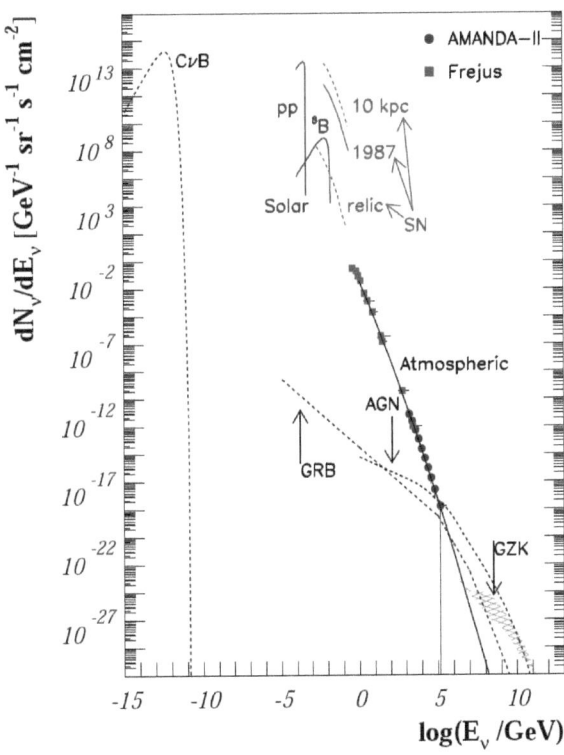

Figure 2.9: *The astrophysical neutrino spectrum including different source predictions. Fluxes from point sources have been scaled by $\frac{1}{4\pi}$ in order to be comparable to diffuse spectra. Figure from [Bec08] after [Rou00]. The atmospheric prediction is averaged over the solid angle is taken from [VZ80], the atmospheric data are from the Fréjus experiment [D+95] (red squares) and from the AMANDA experiment [Mün07, MIC+05](blue circles). The fluxes based on predictions are dashed, solid lines represent those fluxes already measured.*

microwave background. The CνB decoupled in the Early Universe 1 s after the Big Bang, thus the temperature of the blackbody spectrum today corresponds to $\sim 1.9\,\text{K}$ and peaks at milli eVenergies. This background is essentially predicted in the standard model of cosmology but it was yet not possible to test it experimentally due to the small interaction cross section of neutrinos at such small energies.

Source class	Normalization wavelength	ν correlation to wavelength	origin	model(s)
blazars	CRs	proton flux, responsible for $p\gamma$ in the source	jet	[MPR01] [MPE+03]
	> 100 MeV	cascaded π^0 signal connected to $\pi^\pm$ production leading to νs	jet	[MPR01] [MPE+03] [Man95]
	> MeV	cascaded π^0 signal connected to $\pi^\pm$	jet	[Man95] [Ste05]
FSRQs	radio	jet-disk correlation radio power ~ total power	jet	[BBR05a]
FR-II	radio	jet-disk correlation radio power ~ total power	jet	[BBR05b]
radio quiet AGN	X-ray	cascaded π^0 signal	disk	[SS96] [NMB93] [AM04]

Table 2.1: *Neutrino models for Active Galactic Nuclei. Table after [Bec08]. More detailed description see text.*

The sun emits neutrinos in different fusion processes in the MeV range. In the figure neutrinos from the pp fusion chain and from the ^{7}B chain are shown. The neutrino spectrum of SN1987A lies at slightly higher energies, also indicated is the neutrino flux of a supernova in a distance of 10 kpc. A supernova at that distance will be observed by today's neutrino telescopes once it happens. Four orders of magnitudes lower and yet not tested is the flux from SNRs ("relic"). At energies above 0.1 GeV the measured atmospheric neutrino spectrum is indicated measured by Fréjus (red squares) [D+95] and AMANDA (blue circles) [Mün07, MIC+05]. At the highest energies a generic spectrum for GRBs is shown [WB97, WB99] as well as the maximum contribution from AGN [MPR01] and the expected flux from absorbed protons by the GZK effect [YT93]. These neutrinos have not been observed yet due to the high atmospheric background and the limited sensitivity of the detectors. It is the aim of detectors like IceCube and KM3NeT to detect those sources. The following sections will focus on neutrinos from AGN and GRBs, Starburst-Galaxies as possible neutrino sources will be discussed in chapter 3.

2.6.1 AGN as neutrino sources

It is assumed in some interaction models that for each class of AGN the electromagnetic emission is correlated to a neutrino signal. The basic assumptions of the different interaction models are summarized in table 2.1. The models are developed for different AGN classes using different signal hypotheses. The normalization of the neutrino spectrum for each model is done using the either charged cosmic rays or non-thermal photon emission from the given source class. The cosmic ray flux gives evidence for proton acceleration, these accelerated protons interact with the photon field to produce neutrinos. The correlation between neutrinos and photons with MeV to GeV energies can be present if the photons arise from π^0 decays. The π^0 decays imply coincident production of charged pions which decay to leave neutrinos. Since the radio power is connected to the total power via the jet-disk model [FB95, FMB95, FB99] the normalization to the radio signal of AGN can be used. A fraction of the total power of the AGN goes into neutrinos. The X-ray emissions originate at the foot of the jet, in the cited models these X-rays are assumed to be produced in π^0 decays and then cascaded down from TeV energies to X-ray wavelengths in an optically thick environment. For a more comprehensive discussion of AGN neutrino flux models see [Bec07] and [Bec08].

2.6.2 Neutrinos from GRBs

Three phases of non-thermal photon and neutrino emission are expected: precursors in the hours prior to the GRB [RMW03], emission during the prompt phase as well as afterglow emissions [WB00].

Precursor emission

The basic idea of the precursor model developed in [RMW03] is that a shock forms when the pre-GRB matter collides with the wind of the central pulsar or the SNR shell. The shock environment yields a good target for neutrino production by accelerated protons interacting with thermal X-rays in the sub-stellar jet cavity. At this point the burst is still opaque to photon emission. These shocks happen at smaller radii than the prompt emission and at lower boost factors. It is also possible that the neutrino signal could be accompanied by a signal in the far infrared. The low energy part of the neutrino spectrum comes from pp interactions and is proportional to E^{-2}. At energies larger than 10^5 GeV the neutrino flux arises from $p\gamma$ interactions.

Prompt emission

The prompt neutrino emission can be correlated to ultra high energy cosmic rays (UHECRs) since protons are believed to be accelerated in highly relativistic shocks. In turn the acceleration of protons implies neutrino production through photo-hadronic interactions. The neutrino energy spectrum during the prompt photon emission phase in a GRB was determined by

Waxman&Bahcall [WB97, WB99] and can be expressed as

$$\frac{dN_\nu}{dE_\nu} = A_\nu \cdot E_\nu^{-2} \cdot \begin{cases} E_\nu^{-\alpha_\nu+2} \cdot \epsilon_\nu^{b\,\alpha_\nu-\beta_\nu} & ,E_\nu < \epsilon_\nu^b \\ E_\nu^{-\beta_\nu+2} & ,\epsilon_\nu^b < E_\nu \leq \epsilon_\nu^S \\ \epsilon_\nu^S \cdot E_\nu^{-\beta_\nu+1} & ,E_\nu > \epsilon_\nu^S. \end{cases} \qquad (2.3)$$

The spectrum includes two spectral indices, α_ν and β_ν, two break energies, ϵ_ν^b and ϵ_ν^S and a normalization factor A_ν. Their numerical values were determined to

$$\begin{aligned} \alpha_\nu &= 1 \\ \beta_\nu &= 2 \\ \epsilon_\nu^b &\approx 3 \cdot 10^6 \,\mathrm{GeV} \\ \epsilon_\nu^S &\approx 3 \cdot 10^7 \,\mathrm{GeV} \\ A_\nu &\propto d_l^{-2}. \end{aligned}$$

Both, the break energies as well as the spectral indices vary for each individual burst as described in [GHA+04, BSHR06]. The normalization constant A_ν varies for each individual source. It depends on the luminosity distance ($\propto \frac{1}{d_l^2}$) as well as on the fraction of energy transferred into electrons and the fraction of energy transferred into charged pions. In addition, the normalization of the neutrino spectrum scales with the luminosity of the burst. This released energy varies from burst to burst. In addition to this burst-to-burst fluctuation, regular GRBs are distinguished from low-luminosity bursts. Regular, long bursts emit a total isotropic energy of 10^{52} erg for a duration (t_{90}) of the burst of ≈ 10 s. Low-luminosity bursts last longer and and have a lower luminosity. Although only few low-luminosity bursts are observed yet, they are expected to be much more frequent than regular GRBs. For this class, an energy release of $\sim 10^{50}$ erg within around 1000 s is expected. The closest burst observed so far was GRB980425, which was found to be associated with the supernova SN1998bw [G+98]. The host galaxy lies at a redshift of only $z = 0.0085$ (~ 4 Mpc). This burst shows a total energy release of $\sim 10^{47}$ erg, which is an extremely low-luminosity burst. As the luminosity distribution is not well-known at this point, due to low statistics, a fixed value of 10^{51} erg [38] is used in neutrino flux models. An actual burst can be about one order of magnitude more or less luminous.

Afterglow emission

The afterglow emissions are produced when the internal shocks from the original burst hit the interstellar medium producing external shocks [WB00]. The synchrotron emission of electrons gives evidence to relativistic charged particles which implies neutrino production by the baryonic component of the jet and the photon field. The acceleration to such high energies ($E_p > 10^{20}$ eV) implies the neutrino production in proton-photon interactions in environments which

---

[38] 1 erg = 10^{-7} J

are optically thick to proton-photon or proton-proton interactions. In [WB00] Waxman and Bahcall conclude that a significant neutrino flux is directly correlated to the electromagnetic emission during the afterglow phase.

2.6.3 Neutrinos from Starburst-Galaxies

Starburst-Galaxies are galaxies with a high star formation rate and thus also a high supernova rate making them a possible neutrino source. Starburst-Galaxies will be discussed in more detail in the next chapter.

3

Starburst-Galaxies

This thesis focusses mainly on a particular type of galaxy, the Starburst-Galaxy. In the following Starburst-Galaxies will be introduced and then discussed in the context of neutrino astronomy. An analysis with the aim to confine a possible neutrino flux from Starburst-Galaxies will be presented in chapter 5.

Starburst-Galaxies differ from late type galaxies ("normal" galaxies) through their enhanced star formation rate (SFR). An area with higher than average star formation activity is labeled as a *starburst* region. Thus *Starburst-Galaxies* are galaxies which show an enhanced SFR. There is no common classification scheme for Starburst-Galaxies like for the AGN and no uniform criteria that classifies galaxies as Starburst-Galaxies. To classify a galaxy the SFR is evaluated. There are several methods to obtain the SFR in Starburst-Galaxies. The SFR can be calculated by evaluating the HCN and IR emissions from the galaxy. The HCN luminosity is a measure for the density of molecular gas which is necessary to induce star formation. The SFR in dependence of HCN luminosity was determined to [GS04b]

$$SFR_{\rm HCN} \approx 1.8 \cdot 10^7 \left(L_{\rm HCN} \frac{\rm s}{\rm K\,km\,pc^2} \right) \frac{\rm M_\odot}{\rm yr} . \tag{3.1}$$

In [GS04b] the fraction of HCN to CO luminosity $\frac{L_{\rm HCN}}{L_{\rm CO}}$ is proposed as an indicator for Starburst-Galaxies. For galaxies investigated in [GS04b] the authors find $\frac{L_{\rm HCN}}{L_{\rm CO}} \approx 0.1 - 0.25$. The HCN and other molecular luminosities are measured as line emissions at radio wavelengths. The SFR in dependence of the IR luminosity is given by

$$SFR_{\rm IR} \approx 2 \cdot 10^{-10} \left(\frac{L_{\rm IR}}{L_\odot} \right) \frac{\rm M_\odot}{\rm yr} . \tag{3.2}$$

The above relations were obtained using a survey in HCN [GS04a]. The SFR can also be obtained by evaluating measurements in other wavelengths. In [SW09] the SFR was determined from the poly aromatic hydrocarbon (PAH) luminosity which is measured in the mid-infrared at a wavelength of 7.7 μm from the radio luminosity at 1.4 GHz and from far-ultraviolet luminosity for a sample of 287 Starburst-Galaxies also measured in infrared by Spitzer. It was derived that galaxies that were luminous in infrared were dustier. Dust and molecular clouds are necessary for star formation, the molecular clouds are mainly hydrogen clouds which serve as fuel for the star formation. The gas and dust is then heated by the star forming processes and emitting the energy gained as infrared radiation. Starburst-Galaxies emit up to 98% of their energy as infrared light while the Galaxy emits only 30% and the Andromeda galaxy a few percent [CO96]. Thus Starburst-Galaxies are extremely bright in infrared, they are ultra luminous infrared galaxies (ULIRGs) or luminous infrared galaxies (LIRGs). Most of the star formation occurs within about one kiloparsec around the center of the galaxies but there are also Starburst-Galaxies where the star formation occurs throughout the disk. The star formation has to be triggered, i.e. the gas has to be condensed to start the star forming process. The star formation is often triggered by a merging process of two galaxies. In such a process the gas becomes concentrated in the center since stars and gas of the colliding galaxies react differently to the impact of the intruding galaxy. The gas is moving out in front of the stars as they orbit the galactic center. The gravity of these stars pulls back on the gas and the resulting torque on the gas reduces its angular momentum causing it to plunge in the galactic center. While the two galaxies merge more angular momentum is lost and shock fronts then compress and heat the gas, a starburst begins [CO96]. In a typical Starburst-Galaxy between 10 and 300 $M_\odot$ of gas are turned into stars per year. For comparison in the Galaxy there are two to three stars formed each year. This high rate of star formation is typically kept for $10^8 - 10^9$ years. Not all Starburst-Galaxies have their star forming regions in the center, some of them have them spread over the galactic disk. A mechanism which triggers the star formation almost simultaneously in the whole disk is not found yet. An empirical law that connects the SFR to the gas density in a galaxy was found for normal galaxies in 1959 [Sch59] and later extended to star forming galaxies by Kennicutt [Ken98] this expression is known as the *Kennicutt-Schmidt-Law*:

$$SFR \propto \Sigma_{gas}^{1.4} \tag{3.3}$$

Here, Σ_{gas} is the gas density per unit area and SFR the star formation rate per unit area. The exponent 1.4 was determined empirically from observational data.

There can be a connection between the evolution of Starburst-Galaxies and AGN, see e.g. [VBD08, CWY+09]. The central black hole is believed to be fed by star forming activities in the torus of the AGN. Thus some galaxies can be classified as both AGN and Starburst-Galaxy. Recent observations of the Starburst-Galaxies M 82 [A+09f] and NGC 253 [A+09g] in TeV gamma rays ($E_\gamma > 700$ GeV for M 82 and $E_\gamma > 220$ GeV for NGC 253) as well as the observation of both in GeV gamma rays ($E_\gamma > 200$ MeV) by the Fermi LAT [A+09d] yield a strong support to the connection of cosmic ray acceleration and star formation, indicating that supernovae and massive star winds are the dominant accelerators. Both galaxies have the starburst region around their nucleus [VAB96, Ulv00]. No variability in gamma rays was observed which agrees with the emission of the gamma rays from diffuse cosmic ray interactions, though small variations in gamma flux might occur if a supernova recently happened in the galaxy [K+00, B+09b]. Large variations on short timescales would rule out the possibility of the gamma rays being of cosmic ray origin.

A power law behavior was observed between GeV and TeV energies which suggests that a single physical emission mechanism dominates the prodiction of gamma rays at these energies [A+09d].

3.1 A local sample of Starburst-Galaxies

In this section, a sample of local Starburst-Galaxies is presented. The folowing findings as well as the source catalog was already published in [BBDK09]. The data of the individual sources are presented in appendix A. Only local sources are considered, since the aim of this thesis is to investigate the closest sources. The catalog presented here consists of a total of 127 Starburst-Galaxies. This is a sub-sample from a larger sample of 309 Starburst-Galaxies, applying cuts at both FIR and radio wavelengths to ensure a complete, local sample. The initially 309 sources were all classifiied as Starburst-Galaxies earlier [RFG98, CTR05, T+06] and hence, no spectroscopic classification of the sample of starbursts is presented here. The cuts on the initial sample of 309 sources are discussed in the following paragraphs. Also, a small contamination of Seyfert galaxies, as only a small fraction of the sources have X-ray measurements is possible. This contamination is, however, expected to be negligible. Different tests were performed in order to verify that the considered galaxies are indeed starbursts, as presented in the following paragraphs. In order to minimize contamination from Seyfert galaxies, only sources with high ratios of FIR to radio flux density, i.e. $S_{60\mu}/S_{1.4\,\mathrm{GHz}} > 30$ are used. Here, $S_{60\mu}$ is the flux measured in FIR at $60\,\mu$m wavelength and $S_{1.4\,\mathrm{GHz}}$ is the radio flux measured at 1.4 GHz. To test if the sample consists of Starburst-Galaxies as opposed to regular galaxies, the correlation between radio power and FIR luminosity is checked wether it is a direct proportionality. Apart from that, the main criterion for the catalog is that the sources are closer than $z < 0.03$, and that they have both radio and IR detections. The latter gives information about the ratio of the IR to radio signal, which is required to be larger than 30. This ensures a high IR component compared to the radio part, leaving mostly starbursts and only few Seyfert galaxies. Further, sensitivity cuts are applied, sources with a flux density $> 4\,\mathrm{Jy}$ at $60\,\mu$m and a radio flux density at 1.4 GHz larger than 20 mJy are included. Figures 3.1 and 3.2 show the $60\,\mu$m resp. 1.4 GHz luminosity of Starburst-Galaxies versus their luminosity distance. The dotted lines represent the sensitivity for 4 Jy, resp. 20 mJy. Crosses represent all 309 sources selected in the beginning, squares show those 127 sources remaining after the cuts at $S_{1.4\,\mathrm{GHz}} > 20\,\mathrm{mJy}$ and $S_{60\mu} > 4\,\mathrm{Jy}$, as well as $z < 0.03$. Those cuts are applied in order to ensure a complete, local sample in both FIR and radio wavelengths. Since the sources are closer than $z = 0.03$, many of the starbursts are located in the supergalactic plane. Their spatial distribution should therefore be a flat cylinder with a further more spherical component, for those sources not in the supergalactic plane. Therefore it is expected that the number of sources with a flux density larger than S, $N(>S)$, should follow a behavior of $S^{-1} - S^{-1.5}$. A pure S^{-1}−behavior is expected for a flat cylinder, while a spherical distribution results in an $S^{-1.5}$−behavior. Figure 3.3 shows the logarithmic number of sources above an FIR flux density

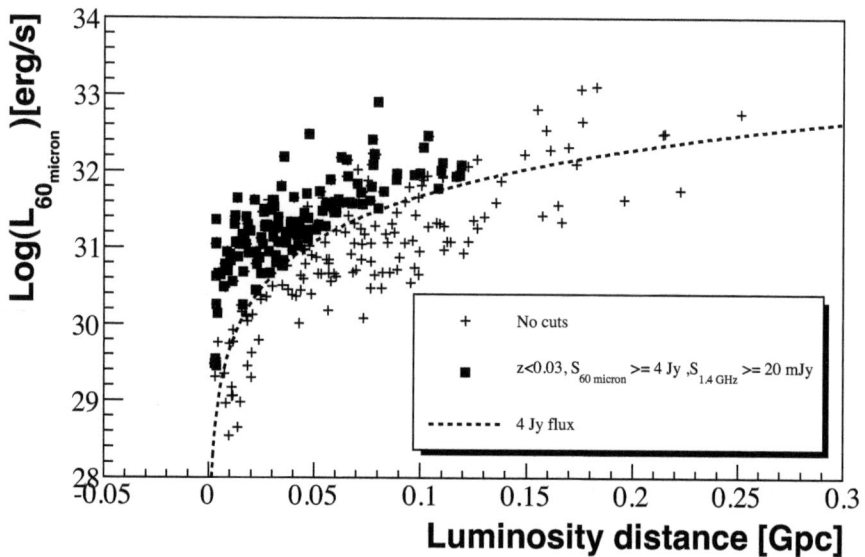

Figure 3.1: 60μ luminosity - distance diagram. Crosses indicate all 309 preselected Starburst-Galaxies, squares show those remaining after the cuts $S_{1.4\,\mathrm{GHz}} > 20\,\mathrm{mJy}$ and $S_{60\mu} > 4\,\mathrm{Jy}$ and $z < 0.03$. The dashed line shows the sensitivity for $S_{60\mu} > 4\,jy$.

$S_{60\mu}$. The data was fitted with the following function:

$$N(> S) = N_0 \cdot (S + S_0)^{-\beta} \qquad (3.4)$$

Here, N_0, S_0 and β are fit parameters. The fit parameters are determined using a log-likelyhood method:

$$\begin{aligned} N_0 &= 3155 \pm 1297.9 \\ S_0 &= (10.56 \pm 3.78)\,\mathrm{Jy} \\ \beta &= 1.2 \pm 0.2 \,. \end{aligned}$$

The behavior $N(> S) \sim S^{-1.2\pm 0.2}$ matches the expectation that the function should lie between $S^{-1.0} - S^{-1.5}$. In the following paragraphs, further investigations are done whether the classification of the 127 sources as starbursts is justified.

3.1. A local sample of Starburst-Galaxies

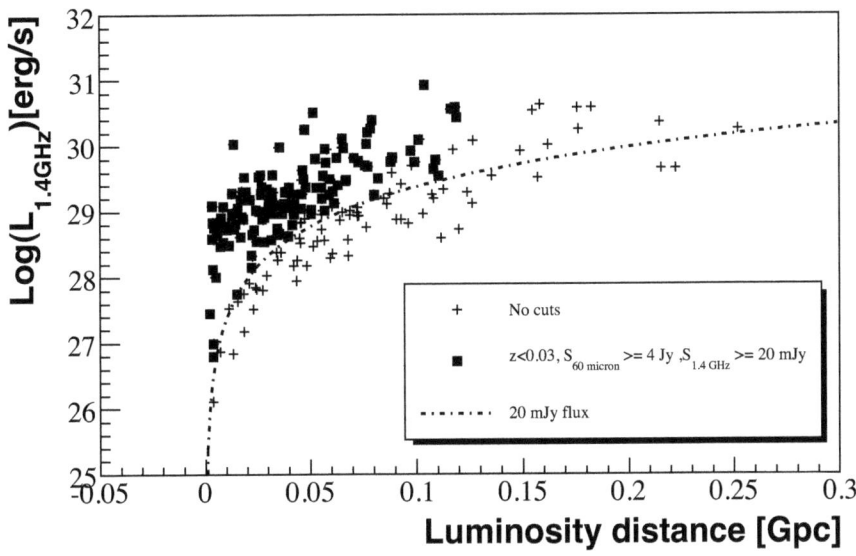

Figure 3.2: 1.4 GHz luminosity - distance diagram. Same notation as Fig. 3.1. The dashed line shows the sensitivity for $S_{1.4\,GHz} > 20\,mJy$.

3.1.1 FIR luminosity versus Radio power

Looking at a well defined sample of galaxies, it turns out that the correlation between radio and far-infrared (FIR) emission is *not* linear, i.e., that the radio luminosity is proportional to the far-infrared luminosity to the power 1.30 ± 0.03 [X+94]. As Xu and collaborators note, the far-infrared emission has two heating sources, stars that later do explode as supernova remnants, and also stars, that will never explode as supernovae. This second population of stars needs to be corrected for, and their contribution to the dust heating needs to be eliminated. This then leads to a corrected far-infrared luminosity, which is directly proportional to the radio luminosity [X+94]. The proportionality holds along a disk in a galaxy, even for fairly short lived phases like a starburst, such as in M 82, and thus requires clearly local physics, with a short readjustment time scale. This poses a severe difficulty for any proposal to explain the radio/FIR correlation. The FIR luminosity in the range of 60 μm and 100 μm is given as [X+94]:

$$L_{\rm FIR} := 4\pi\, d_l^2 \cdot F_{\rm FIR}\,. \tag{3.5}$$

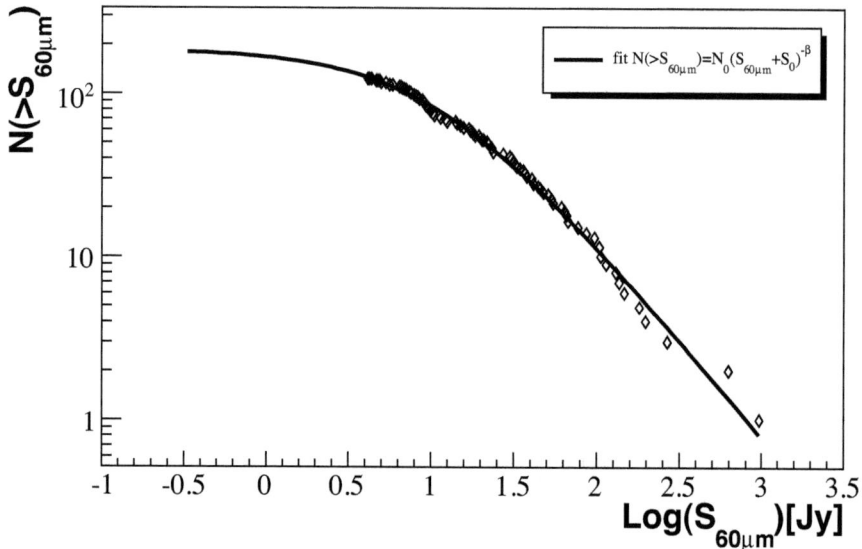

Figure 3.3: $\log N - \log S$ representation of the catalog. An $S^{-1.2}$–fit matches the data nicely, with a turnover at $S_0 = 10.56$ Jy.

Here, d_l is the luminosity distance of the individual sources and

$$F_{\mathrm{FIR}} := 1.26 \cdot 10^{-14} \cdot \left[2.58 \cdot \left(\frac{S_{60\mu}}{\mathrm{Jy}} \right) + \left(\frac{S_{100\mu}}{\mathrm{Jy}} \right) \right] \; \mathrm{W\,m^{-2}} \qquad (3.6)$$

is the FIR flux density at Earth as defined in [H+88]. The normalization factor comes from the frequency integration and from the conversion of Jy to W/m²/Hz. In Fig. 3.4, the logarithm of the radio power at 1.4 GHz, $P_{1.4\,\mathrm{GHz}}$ versus the logarithm of the FIR luminosity L_{FIR} is shown for the catalog. The circles show the single sources and the solid line is a fit through the data. The fit yields a correlation of

$$P_{1.4\,\mathrm{GHz}} \propto L_{\mathrm{FIR}}^{1.1}. \qquad (3.7)$$

As only 11 out of the 127 radio measurements have errors, it is difficult to estimate the exact error on this result. If the average of 4.1% uncertainty, which was derived from the known 11 errors, is used an uncertainty to the spectral index of around 10% is obtained. Therefore, it is expected that the majority of sources in the sample are starbursts.

3.1. A local sample of Starburst-Galaxies

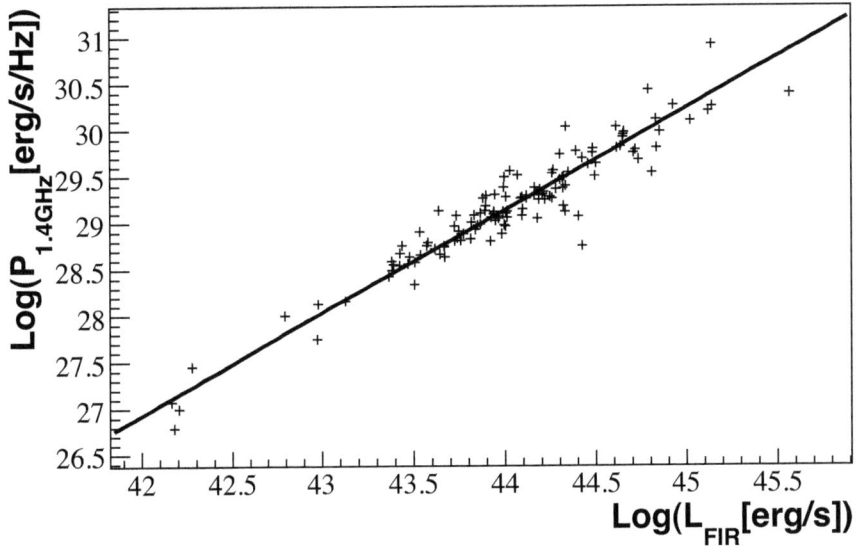

Figure 3.4: Radio power $P_{1.4\,\mathrm{GHz}}$ at $\nu = 1.4\,\mathrm{GHz}$ versus FIR luminosity L_{FIR}. A proportionality of $P_{1.4\,\mathrm{GHz}} \propto L_{FIR}^{1.1}$ is found.

3.1.2 Infrared to radio flux density ratio

Generally, regular galaxies are distinguished from active galaxies by their ratio of the FIR flux density at 60 µm, $S_{60\mu}$, and the radio flux density at 1.4 GHz, $S_{1.4\,\mathrm{GHz}}$:

$$s_{60\mu/1.4\,\mathrm{GHz}} := \frac{S_{60\mu}}{S_{1.4\,\mathrm{GHz}}}. \qquad (3.8)$$

For Seyfert galaxies, this ratio is about $s_{60\mu/1.4\,\mathrm{GHz}} \sim 10$, while it is significantly higher in the case of starburst galaxies, $s_{60\mu/1.4\,\mathrm{GHz}} \sim 300$. While most of the Seyfert galaxies are removed from the sample using this method, a small fraction can still be present. The histogram of the ratio between the FIR flux density at 60 µm and the radio flux density at 1.4 GHz is shown in Fig. 3.5. All 127 sources have a ratio of $s_{60\mu/1.4\,\mathrm{GHz}} > 30$, which confirms that the majority of the sources are not likely to be Seyferts, leaving behind only a small fraction of potential Seyferts.

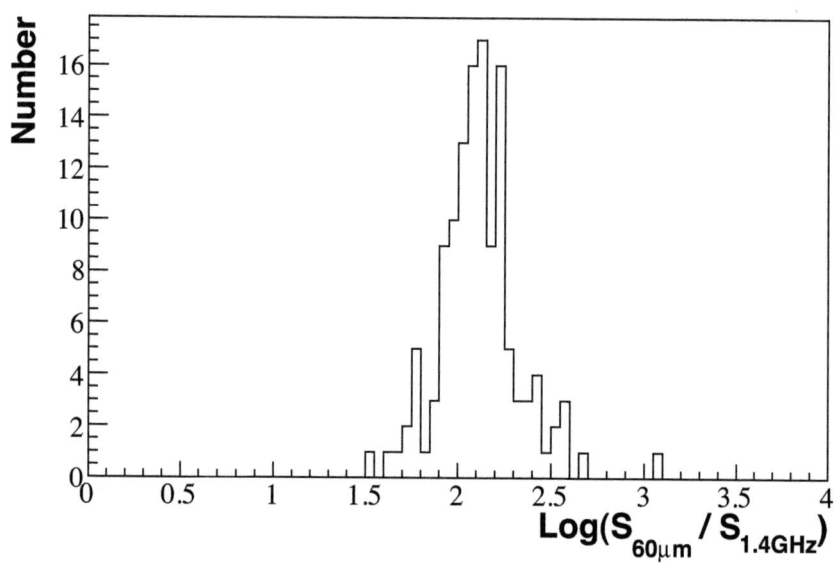

Figure 3.5: Ratio of the flux density at 60 μm and at 1.4 GHz. All sources in the sample have ratios larger than 30, which indicates a high star formation rate. The median is around 100. This matches previous investigations, e.g. [BEW85], who find a mean value of 250 at higher radio frequencies, $\nu = 5$ GHz.

3.1.3 Radio to Infrared and X-ray to Infrared spectral indices

A further criterion of distinguishing regular galaxies and Seyferts is their spectral index from X-ray to IR (XIR) and from radio to IR (RIR). The diagram of the XIR (1 keV to 60 μm) versus RIR (5 GHz to 60 μm) index of the sources is shown in Fig. 3.6. Derived from figure 3 in [R+93], Starburst-Galaxies have spectral indices scattering around $(RIR, XIR)_{starburst} \sim (0.6, -1.9)$, Seyfert I galaxies show $(RIR, XIR)_{Sy-I} \sim (0.48, -1.2)$, Seyfert II galaxies have $(RIR, XIR)_{Sy-II} \sim (0.47, -1.6)$ and quasars are located at $(RIR, XIR)_{quasar} \sim (0.28, -1.1)$.
The values for the RIR and XIR indices of Starburst-Galaxies given by [CKB89] are slightly higher, which matches the sample examined here: [CKB89] give a RIR index of 0.82 and a XIR index of -1.66. The sample presented here shows average values of
$(\overline{RIR}, \overline{XIR}) = (0.82 \pm 0.01, -1.77 \pm 0.021)$ which is compatible with the expected result. Still, no X-ray data for all the sources is availiable, so there may still be some contamination from both Seyferts and regular galaxies in the sample, which is expected to be negligible.

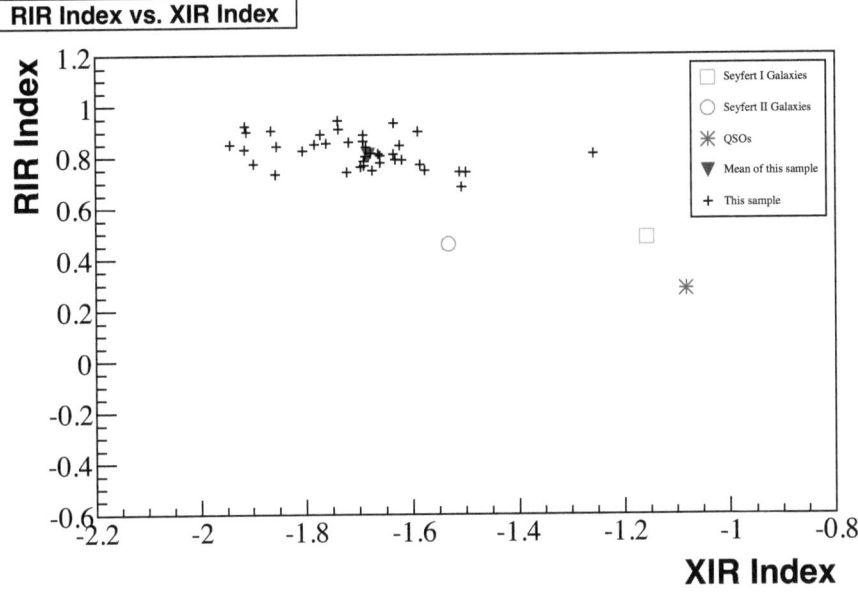

Figure 3.6: *Radio-to-IR spectral index versus X-ray-to-IR spectral index. The crosses represent those 48 sources in the catalog with radio, FIR and X-ray measurements. The triangle shows the average of the values. The open circle shows the average location of Seyfert I galaxies, the open square represents average Seyfert II galaxies and the star indicates QSOs. The last three values are taken from [R+93]. Note that individual galaxies scatter around the given values [CKB89].*

3.2 Starburst-Galaxies as neutrino sources

As stated in [BBDK09], the cosmic ray intensity from Starburst-Galaxies scales with the radio and infrared emission of the sources. There are two source classes within starbursts that can accelerate cosmic rays to high energies, namely shock fronts of supernova remnants and long Gamma Ray Bursts (GRBs), both discussed in section 2.4. In the first case, maximum energies are limited to less than 10^{15} eV and thus, the cosmic rays from starbursts cannot be observed directly due to the high cosmic ray background in our own Galaxy. Gamma Ray Bursts, on the other hand, were proposed as the origin of cosmic rays above the ankle, i.e. $E_{CR} > 3 \cdot 10^{18}$ eV, see [Vie95, Wax95]. Since a high star formation rate as it is present in Starburst-Galaxies, leads to a high rate of supernova explosions, an enhanced rate of long GRBs is expected. Thus, for closeby sources, the distribution of Starburst-Galaxies can be used to test the hypothesis of cosmic rays from starbursts, as also discussed in [B+09a]. The diffuse high energy neutrino flux from Starburst-Galaxies can be calculated using the supernova rate as done in [BBDK09]. In this calculation the SNR are expected to accelerate particles up to energies below 10^{15} eV since SNR in the Galaxy do the same. The calculated neutrino spectrum shows a cut off at about 10^{15} eV since protons are not expected to be accelerated beyond this energy in SNR. The calculated diffuse neutrino flux from Starburst-Galaxies is shown in figure 3.7 and it is out of reach for neutrino detectors like AMANDA and IceCube. For a description of these detectors see section 4. Previously a flux prediction by Loeb and Waxman was given [LW06] which derived the neutrino flux from the FIR luminosity assuming that 100% of the diffuse FIR background derives from Starburst-Galaxies. This yields a quite high flux prediction which was doubted by Stecker [Ste07]. He pointed out that only 23% of the FIR background originates from Starburst-Galaxies.

3.2.1 Gamma Ray Bursts and starbursts

Starburst-Galaxies show an enhanced rate of supernova explosions due to their large star formation rate. Thus an increasing rate of long Gamma Ray Bursts (GRBs) directly linked to SN-Ic events [M+03b] is expected. If long GRBs are the dominant sources of UHECRs, the contribution from nearby objects should follow the distribution of Starburst-Galaxies. In the following calculations, it is assumed that every SN-Ic explosion is accompanied by a particle jet along the former star's rotation axis, i.e. by a GRB. The opening angle of the GRB jet, θ determines, how many SN-Ic can be observed as GRBs, see e.g. [B+03, R+08].

$$\dot{n}_{\text{GRB}} = \epsilon \cdot \dot{n}_{\text{SN-Ic}}. \tag{3.9}$$

Here, $\dot{n}_{\text{GRB}}$ is the GRB rate in a galaxy and $\epsilon = (1 - \cos(\theta))$ is the fraction of SN-Ic which can be seen as GRB. The opening angle is expected to be less than $\sim 10°$ for the prompt emission. Afterglow emissions and precursors can have a larger opening angles [M+08]. Putting the focus on prompt emission, an optimistic opening angle of $\sim 10°$ is used, yielding

$$\epsilon = 0.015. \tag{3.10}$$

Further, observational data show that core collapse supernovae of type Ic contribute with 11% to the total supernova rate in starbursts [CT01]. Using equation 3.9 the GRB rate in a starburst galaxy is directly correlated to the supernova rate $\dot{n}_{SN}$,

$$\dot{n}_{\text{GRB}} = \epsilon \cdot \xi \cdot \dot{n}_{\text{SN}} \tag{3.11}$$

with $\xi \sim 0.11$ as the fraction of heavy SN among all SN. The supernova rate is correlated with the FIR luminosity of the galaxy [M+03a],

$$\dot{n}_{\text{SN}} = (2.4 \pm 0.1) \cdot 10^{-12} \cdot \left(\frac{L_{\text{FIR}}}{L_\odot}\right) \text{ yr}^{-1}. \tag{3.12}$$

3.2. Starburst-Galaxies as neutrino sources

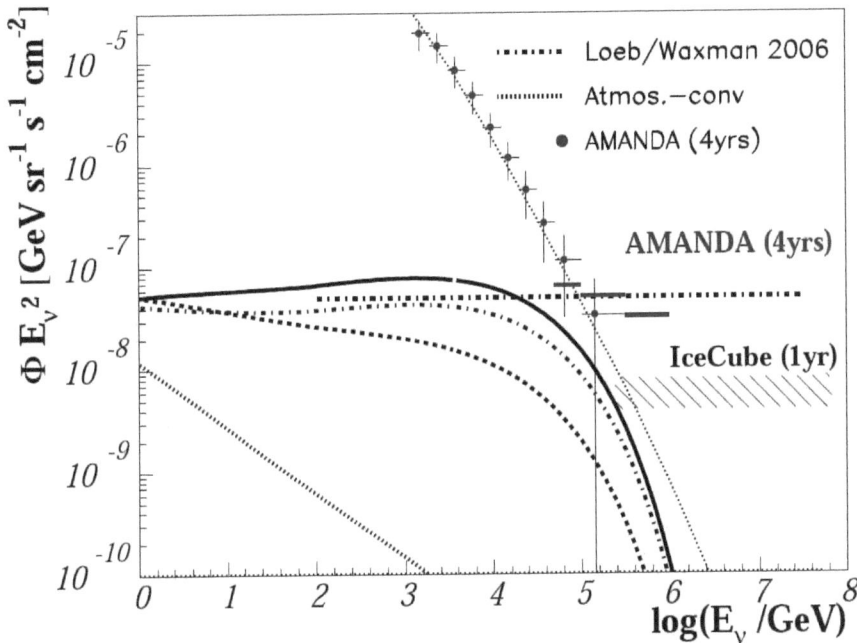

Figure 3.7: *Expected diffuse high-energy neutrino flux from SNRs in Starburst-Galaxies. Data points show the atmospheric neutrino background as measured by the AMANDA experiment (data between 2000 and 2003) [M+07, Mün07]. The prediction of atmospheric neutrinos is taken from [Vol80]. AMANDA limits are for the same data sample, derived from the fact that no significant excess above the atmospheric background was observed. The dot-dashed line shows the prediction by [LW06]. Figure taken from [BBDK09].*

The FIR luminosity is expressed in terms of the solar luminosity $L_\odot = 3.839 \cdot 10^{33}$ ergs and is given in the range of 60 μm and 100 μm by [X+94]

$$L_{\text{FIR}} = 4\pi d_l^2 \cdot F_{\text{FIR}}. \tag{3.13}$$

Here, d_l is the luminosity distance of the individual source and

$$F_{\text{FIR}} = 1.26 \cdot 10^{-14} \cdot \left(2.58 \cdot S_{60/\mu} + S_{100\mu}\right) \, \text{W m}^2 \tag{3.14}$$

is the FIR flux density as defined in [H+88]. $S_{60\mu}$ and $S_{100\mu}$ are the measured flux densities at 60 μm and 100 μm, both measured in Jy. Using equation 3.12 to determine the supernova rate, equation 3.11 yields a rate of

$$\dot{n}_{\text{GRB}} = 3.8 \cdot 10^{-15} \cdot \left(\frac{L_{\text{FIR}}}{L_\odot}\right) \cdot \left(\frac{\epsilon}{0.015}\right) \cdot \left(\frac{\xi}{0.11}\right) \text{ yr}^{-1} \quad (3.15)$$

per starburst. For a $1\,\text{km}^3$ neutrino detector with a lifetime of 10 years (like IceCube or KM3NeT) luminosities of around $3 \cdot 10^{13} \cdot L_\odot \sim 10^{47}$ ergs are required for a single event within these 10 years. None of the sources in the catalog provides such a high luminosity. However, if a larger number of starbursts is considered for an analysis, the total luminosity increases and so does the probability of observing a GRB. Figure 3.8 shows the total GRB rate for a number of $N_{\text{starbursts}}$ galaxies,

$$\dot{n}_{\text{GRB}}^{\text{tot}}(N_{\text{starbursts}}) = \sum_{i=1}^{N_{\text{starbursts}}} \dot{n}_{\text{GRB}}(i\text{th starburst}). \quad (3.16)$$

In the figure the GRB rates achieved in the single starbursts are summed up, starting with the most luminous source, adding sources in descending luminosity order. The points show the GRB rate summing up over all starbursts in the sample, starting with the strongest one and adding the next strongest sources subsequently. On total, 0.03 GRBs per year are expected to be observable in the sample. The squares display the total GRB rate, summing up sources in the northern hemisphere, which corresponds to IceCube's Field of View (FoV). Here, 0.02 GRBs per year are expected. This number can be enhanced significantly when taking weaker sources into account which were not included in the sample in order to ensure completeness, see section 3.1.

3.3 Enhanced neutrino flux from GRBs in starbursts

Due to the high atmospheric background seen by high-energy neutrino telescopes, the detection of a diffuse neutrino signal from GRBs in nearby starbursts will not be possible. However, with a timing analysis one might be able to identify Gamma Ray Bursts in neutrinos. In such an analysis, the location of a nearby starburst can be chosen as a potential neutrino hot-spot. By selecting a time window of the typical duration of a long GRB ($\sim 100\,\text{s}$) the atmospheric background can then be reduced to close to zero. In this context the general neutrino intensity and in particular the possibility of neutrino detection with IceCube are discussed.

For the first time the neutrino energy spectrum during the prompt photon emission phase in a GRB was determined by Waxman&Bahcall [WB97, WB99] and can be expressed as

$$\frac{dN_\nu}{dE_\nu} = A_\nu \cdot E_\nu^{-2} \cdot \begin{cases} E_\nu^{-\alpha_\nu+2} \cdot \epsilon_\nu^{b\,\alpha_\nu-\beta_\nu} & , E_\nu < \epsilon_\nu^b \\ E_\nu^{-\beta_\nu+2} & , \epsilon_\nu^b < E_\nu \leq \epsilon_\nu^S \\ \epsilon_\nu^S \cdot E_\nu^{-\beta_\nu+1} & , E_\nu > \epsilon_\nu^S. \end{cases} \quad (3.17)$$

The spectrum includes two spectral indices, α_ν and β_ν, two break energies, ϵ_ν^b and ϵ_ν^S and a normalization factor A_ν. For the GRBs in starbursts, these parameters were discussed in detail

3.3. Enhanced neutrino flux from GRBs in starbursts

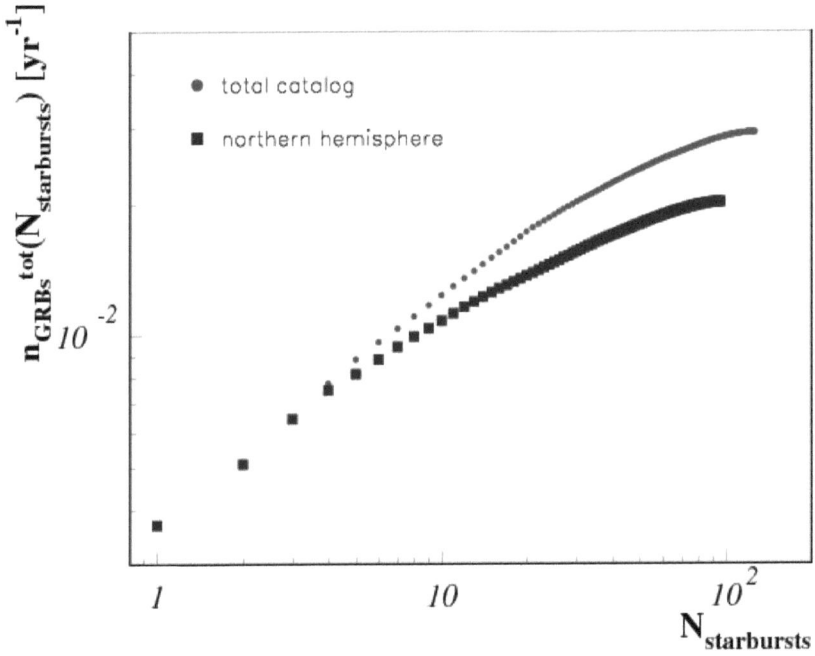

Figure 3.8: Number of GRBs per year in the starburst catalog, including $N_\text{starbursts}$ sources, starting with the strongest one. The total GRB rate in the sample, including all 127 sources is $0.03\,\text{yr}^{-1}$, this means that a GRB could be observed every 30 years on average. The total GRB rate just in the northern hemisphere is $0.02\,\text{yr}^{-1}$ or an occurrence every 50 years, the data shown in squares. These source lie in IceCube's FoV.

in [BBDK09]. Their numerical values were determined to

$$\alpha_\nu = 1$$
$$\beta_\nu = 2$$
$$\epsilon_\nu^b \approx 3 \cdot 10^6 \,\text{GeV}$$
$$\epsilon_\nu^S \approx 3 \cdot 10^7 \,\text{GeV}$$
$$A_\nu \propto d_l^{-2}.$$

The normalization constant A_ν is calculated for each individual source. It depends on d_l^{-2} as well as on the fraction of energy transferred into electrons and the fraction of energy transferred into charged pions. In addition, the normalization of the neutrino spectrum scales with the luminosity of the burst. This released energy varies from burst to burst. In addition to this burst-to-burst fluctuation, regular GRBs are distinguished from low-luminosity bursts. Regular, long bursts emit a total isotropic energy of 10^{52} erg for a duration (t_{90}) of the burst of ≈ 10 s. Low-luminosity bursts last longer and and have a lower luminosity. Although only few low-luminosity bursts are observed yet, they are expected to be much more frequent than regular GRBs. For this class, an energy release of $\sim 10^{50}$ erg within around 1000 s is expected. The closest burst observed so far was GRB980425, which was found to be associated with the supernova SN1998bw [G+98]. The host galaxy lies at a redshift of only $z = 0.0085$. This burst shows a total energy release of $\sim 10^{47}$ erg, which is an extremely low-luminosity burst. As the luminosity distribution is not well-known at this point, due to low statistics, a fixed value of 10^{51} erg was used. An actual burst can be about one order of magnitude more or less luminous. Now, to estimate the neutrino flux from a standard GRB for a single starburst in the sample, dependent on the distance of the starburst, the normalization is calculated. The other parameters are kept constant and hence the results can only serve as a rough estimate. Both the break energies as well as the spectral indices vary for each individual burst as described in [G+04].

Expected event rates in IceCube

As shown in figure 3.8, 0.02 GRBs per year are expected to occur in the 96 starbursts in the sample in the northern hemisphere, IceCube's FoV. However, this rate can be enhanced if all sources in the northern hemisphere would be taken into account. For completeness reasons only the brightest ones were considered here. This enhances the possibility to detect a GRB from a starburst in the super galactic plane within the lifetime (10 years) of IceCube. The prospects for KM3NeT are slightly worse, since only 0.01 GRB per year is expected in the southern hemisphere, where KM3NeT's FoV will be focused. The number of events per GRB expected in IceCube can be calculated by folding IceCube's effective area A_eff [Mon08] with the GRB spectrum dN_ν/dE_ν

$$N_\text{events} = \int_{E_\text{th}}^{\infty} A_\text{eff}(E_\nu) \cdot \frac{dN_\nu}{dE_\nu} \, dE_\nu. \tag{3.18}$$

Here, the weak dependence of the A_eff from the declination of the burst is neglected. For the threshold energy $E_\text{th} = 100\,\text{GeV}$ is used. This is the general detection threshold of IceCube [A+04]. Since events can be selected in a small time window with the typical duration of a long GRB, $10 - 100$ s, the atmospheric background can be reduced close to zero. Figure 3.9 shows the histogram of the numbers of events expected in IceCube from an average GRB with an isotropic energy of $E_\gamma^\text{iso} = 10^{51}$ erg from the 96 starbursts in the sample. These numbers range from 0.1 to about 300 events per burst, depending on the distance of its host galaxy. These numbers lie between 1 and 5 orders of magnitude above the numbers of events for GRBs

3.3. Enhanced neutrino flux from GRBs in starbursts

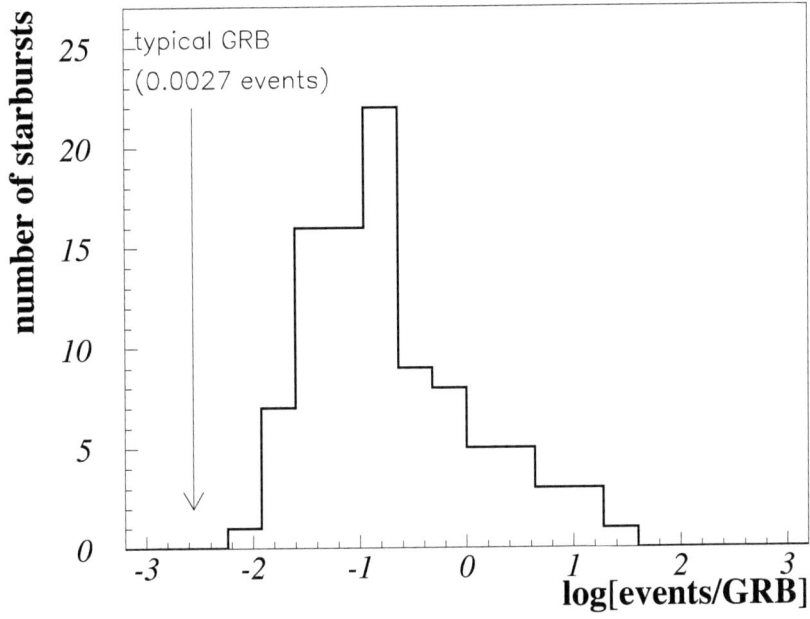

Figure 3.9: *Histogram of the number of events in IceCube from the 96 starburst galaxies in the northern hemisphere. Depending on the distance of the starburst, a burst would result in between 0.1 and several 100 events in IceCube in a small time window of 10 − 100 seconds. A regular burst at $z \sim 1-2$ gives only 0.027 events as indicated in the figure. The main reason for the increased signal is that the bursts would come from starbursts closer than $z = 0.03$.*

typically observed by satellite experiments like Swift, BATSE and Fermi. If such GRB occurs in the sample of Starburst-Galaxies presented here it is a unique opportunity to study the hadronic component of GRBs.

4

Neutrino telescopes at the South Pole

In this chapter the two neutrino detectors located at the South Pole, AMANDA[1] and IceCube are described. As in this thesis, data from both AMANDA and IceCube have been used to perform analyses on the search of neutrinos from point source classes. Currently operational is IceCube, the AMANDA detector was the predecessor of IceCube and has been decommissioned. Both detectors use the Čerenkov light produced by secondary particles produced in neutrino interactions in the ice. Neutrino detection using the Čerenkov effect was explained in section 2.5.4. The infrastructure for building and running experiments like AMANDA and IceCube at the South Pole is provided by the Amundsen-Scott South Pole station, this station provides power for the detectors and habitation for the personnel.

4.1 The AMANDA telescope

AMANDA was the first neutrino detector to be built in ice. AMANDA consists of 677 photomultipliers (PMTs) which are situated between 1500 m and 2000 m deep in the antarctic ice shield. Each PMT is housed in a glass sphere as a pressure housing together with a voltage divider. The PMTs contained in the glass spheres are called optical modules (OMs). The OMs are used to detect Čerenkov radiation produced by secondary particles which were produced in neutrino interactions in the ice. The PMTs in the OMs are looking downwards because it is aimed to detect light from upward going particles. Upward going particles were most likely produced by a neutrino from an extra terrestrial source since neutrinos can travel through the Earth. Downward going particles are mainly produced in the Earth's atmosphere. The OMs are connected to steel cables, each holding between 10 and 42 OMs. These steel cables with the OMs are called strings and AMANDA consists of 19 strings. Along each string a cable for

---

[1] Antarctic Muon And Neutrino Detector Array

the high voltage supply and the signal transmission leads from each OM into the Martin A. Pomerantz Observatory (MAPO) which houses the data acquisition systems and high voltage supplies for AMANDA. AMANDA has been built from 1993 until 2000. In its final stage the detector is called AMANDA-II.

To connect the OMs to the MAPO electrical cables as well as optical fibers are used. In an electrical channel a pulse widens up from its typical length of ~ 20 ns to ~ 200 ns. In an optical channel the pulse keeps its length of ~ 20 ns since there is almost no dispersion in optical fibers. AMANDA delivered valuable results such as flux limits on the neutrino flux from point sources [A+09b] and the atmospheric neutrino flux [Mün07, M+07]. Although AMANDA would be still operable it was decided to shut it down to conserve electrical power which is quite limited at the South Pole. AMANDA was decommissioned on May 11th 2009 after eight years of operation in its final configuration. The success of AMANDA paved the way for building an even larger neutrino telescope at the South Pole, the IceCube detector which will instrument about a km^3 of ice. IceCube will be described in the next section.

4.2 The IceCube neutrino telescope

The IceCube detector is the larger successor of AMANDA. It uses the same detection technique and is designed to detect neutrinos in an energy range from initially ~ 100 GeV to 1 EeV. IceCube has three subdetectors, the part inside the ice called InIce, an airshower array at the surface called IceTop and a low energy extension in the ice called DeepCore. IceCube is currently under construction, each of the subdetecors will be explained in the following. Figure 4.1 shows IceCube in its final configuration.

4.2.1 The InIce detector

The InIce detector of IceCube in its final stage is planned to consist of 5160 digital optical modules (DOMs), an improved version of the AMANDA OMs, deployed between 1450 m and 2450 m in the antarctic ice [Kar08]. The main difference between the AMANDA OMs and the IceCube DOMs is that in the IceCube DOMs the PMT signals are digitized in and then sent to the surface. Transmitting the PMT pulses in digital form the pulses do not suffer from dispersion like it was the case with AMANDA. The signals are sent through twisted pair cables to the surface. During construction the detector is already taking data and analyses are performed with this data. The first string of IceCube was deployed in the austral summer 2004/2005, another eight strings one year later forming the 9-string detector (IC-9). In the season 2006/2007 13 more strings were deployed making up the 22-string detector (IC-22), the data from this detector configuration was used in this thesis. The next stage of the detector was IC-40 with the installation of another 18 strings in the summer 2007/2008. In the latest season, 2008/2009 19 strings (including one string for DeepCore) were deployed extending the

4.2. The IceCube neutrino telescope

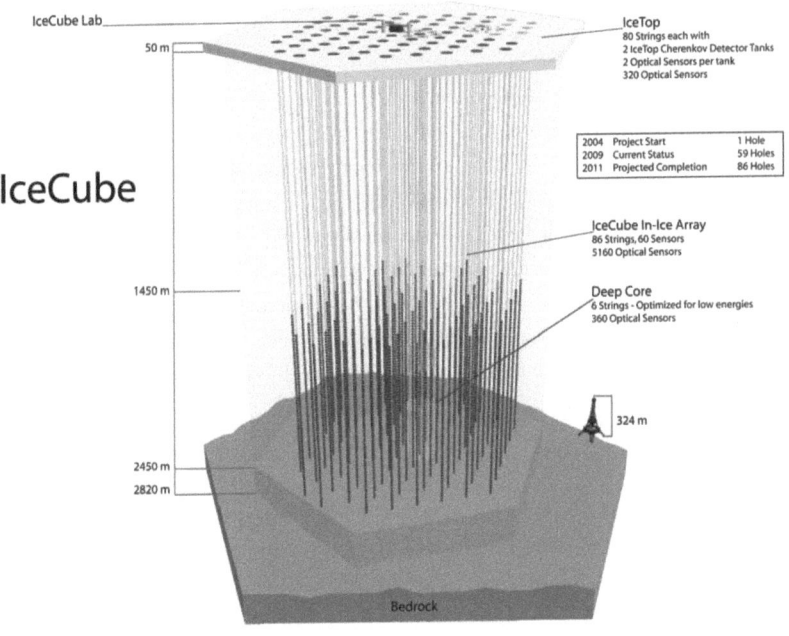

Figure 4.1: *The IceCube detector including InIce, IceTop and DeepCore [Col09].*

Year	Detector	Strings deployed	Total strings
2004/2005	String-21	1	1
2005/2006	IC-9	8	9
2006/2007	IC-22	13	22
2007/2008	IC-40	18	40
2008/2009	IC-59	19	59

Table 4.1: *The different stages completed so far during construction of the InIce detector of IceCube.*

detector to 59 strings in total. This is the configuration IceCube is currently taking data with. The stages of construction are summarized in table 4.1. The current season, 2009/2010, is ongoing with another 5 strings deployed as of December 2009 aiming for a total of 18 strings to be deployed this season. The detector is due for completion in 2010/2011, the InIce detector will have 86 strings according to current plans.

4.2.2 The IceTop detector

Placed on the surface above the InIce detector there is an air shower array labeled IceTop. IceTop measures cosmic ray air showers and will consist of 160 tanks at the surface with two DOMs each when it will be finished. Currently IceTop consists of 59 tanks. IceTop can serve as a veto and a calibration for the InIce detector and can measure the energy spectrum and study the chemical composition of the cosmic rays. IceTop measures the energy deposited in the array and the lateral distribution of the shower signal. These values are necessary to study the composition, heavier nuclei have more energy and a flatter lateral distribution. In addition to this IceTop also measures the angular distribution of the showers and the angular distribution of muon bundles as well as the altitude at which the shower has its maximum. These parameters are also sensitive to the chemical composition of the cosmic rays [Sta09].

4.2.3 The DeepCore extension

Until the decommissioning of AMANDA, AMANDA was integrated into IceCube for detecting low energy events. The InIce detector measures at slightly higher energies as AMANDA ($E > 100\,\text{GeV}$ for IceCube compared to $E > 50\,\text{GeV}$ for AMANDA) since the spacing between the AMANDA OMs is closer than the spacing between the IceCube DOMs. The first DeepCore string was deployed in January 2009, the complete DeepCore detector consisting of 6 strings will be deployed early 2010. With the DeepCore extension the lower energy threshold of IceCube is expected to be an order of magnitude lower, below $10\,\text{GeV}$. Each string of DeepCore holds 60 DOMs, 10 DOMs on each string at shallow depths between $1750\,\text{m}$ and $1850\,\text{m}$ used as a veto for the remaining 50 DOMs at depths between $2100\,\text{m}$ and $2450\,\text{m}$ [Wie09]. The deep ice is on average twice as clear as the shallow ice which is an advantage of DeepCore compared to AMANDA which was at shallow depths.

5

A stacking analysis with AMANDA-II and IC-22 data

In this chapter a stacking analysis for different potential neutrino point source classes is presented. It has previously been shown that the source stacking method yields large improvements in sensitivity in an analysis of generic AGN classes using four years [Gro06] and five years [A+07] of AMANDA-II data. In this thesis this analysis was redone using two new data samples, one covering seven years of AMANDA-II data and one covering 275.7 days of IC-22 data. In addition improved source catalogs were defined, a catalog of Starburst-Galaxies, see section 3.1, as well as source catalogs for blazars, Flat Spectrum Radio Quasars and Pulsars detected by the Fermi LAT [Abd09]. In the following sections the stacking technique which is sensitive to a cumulative signal of a generic source class will be introduced, the stacked source classes as well as the data samples will be presented and the results will be shown.

5.1 The source stacking method

The source stacking method has successfully been used in optical as well as radio astronomy. It has been applied the first time in neutrino astronomy in [Gro06]. The present analysis uses the same technique. The source stacking method is sensitive to a cumulative signal of a generic source class. The source classes themselves have to be defined accurately beforehand according to a well defined signal hypothesis. Such a hypothesis is typically the correlation of a measured photon flux at a certain wavelength band to be connected to the neutrino flux. Proton acceleration is expected at the same site and neutrinos are produced when the accelerated protons interact with the ambient matter or photon field. Thus for the AGN it can be assumed that the neutrino signal is correlated to the radio signal. The signal hypothesis for the Starburst-Galaxies is that the neutrinos are produced in SNRs inside the Starburst-Galaxies which are connected to the star forming activity. As a measure for the star forming activity serves the FIR flux at 60 μm which is then assumed to be correlated to the neutrino flux. Stacking uses the fact that signal and background grow in different speeds when added for multiple sources. While the signal grows linearly with the number N of sources the background grows proportional to $\frac{1}{\sqrt{N}}$ assuming Gaussian statistics. Stacking can therefore suppress background and hence increase the signal sensitivity. Signal and background are evaluated for a certain angular search bin where its size has to be determined for each source class. Both, the number of sources to be stacked N, as well as the angular radius of the search bin r_{bin} are parameters of the analysis which have to be optimized using Monte Carlo simulations. The sources are sorted according to the signal hypothesis as described above. The background expectation derives from the number of events inside the zenith band (off source) of the sources normalized to the area of the search bin

$$n_{bg} = \frac{A_{bin}}{A_{zb}} \cdot n_{zb}. \tag{5.1}$$

Here, A_{bin} denotes the area of the search bin, A_{zb} the area of the according zenith band, n_{bin} and n_{zb} the event numbers in the search bin and the zenith band. The areas of the search bin and the zenith band are defined as

$$A_{bin} = 2\pi \cdot (1 - \cos(r_{bin})) \tag{5.2}$$

and

$$A_{zb} = 2\pi \cdot (\sin(\delta_{min}) - \sin(\delta_{max})). \tag{5.3}$$

In these equations r_{bin} is the angular radius of the search bin and δ_{min} and δ_{max} the minimum and maximum declination angles limiting the zenith band. Astronomical coordinates use the declination (δ or DEC) and right ascension (α or RA), on a technical level the detector coordinates zenith angle Θ and azimuth angle Φ are used. At the South Pole the relation between declination and zenith is $\delta = \Phi + 90°$, the conversion from right ascension and azimuth depends on the time of the event and is done with the analysis software. Figure 5.1 illustrates the search bin and the zenith band.

To justify that the background estimation can be taken from the events in a zenith band, the event distribution in right ascension has to be flat. Both distributions are shown in figure 5.2, for AMANDA on the left and for IC-22 on the right. The distributions are flat within their error bars where the error is the square root of the number of events in each bin. In the stacking approach the signal (on source) and background (off source) events are summed up for N sources

$$SIG(N) = \sum_{i=1}^{N} SIG_i \tag{5.4}$$

$$BG(N) = \sum_{i=1}^{N} BG_i - BG_i^{overlap}. \tag{5.5}$$

Here, SIG_i and BG_i are the number of signal and background events for the ith source. The quantity $BG_i^{overlap}$ is the number of events in the overlap region if two sources have overlapping

5.2. The source samples

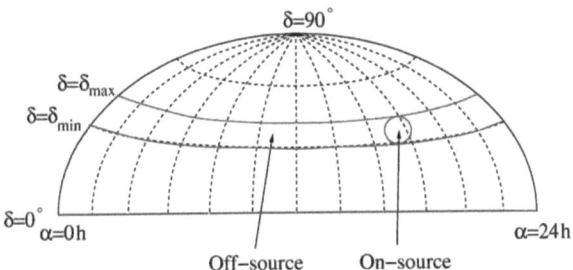

Figure 5.1: *The search bin (on source) and the associated zenith band (off source) enclosed by δ_{min} and δ_{max}. Figure from [Gro06].*

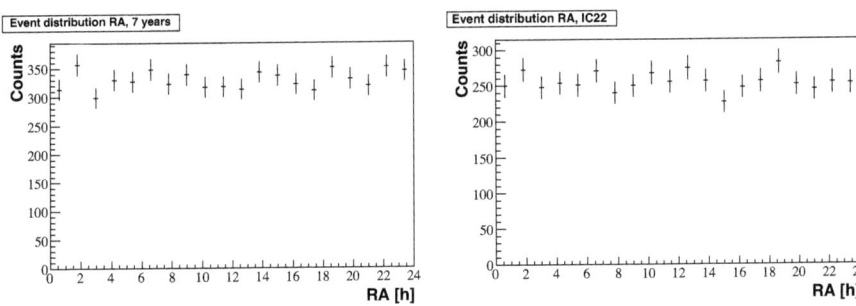

(a) The RA distribution for the AMANDA data set   (b) The RA distribution for the IC-22 data set

Figure 5.2: *Right ascension distribution for the AMANDA and IC-22 data, both distributions are flat.*

zenith bands. This number is obtained analogously to equation 5.1 by replacing A_{bin} with the space angle of the overlap region.

5.2 The source samples

For the analyses performed here only sources in the northern hemisphere which are not in the galactic plane ($|b| \geq 10°$) were selected. The cut on the galactic plane was done to exclude possible galactic sources. An exception to this criteria is the sample of pulsars. As pulsars being of galactic origin, also sources within the galactic plane were selected. In this theses seven generic source catalogs are used for the two stacking analyses:

- **Starburst-Galaxies**
  The source catalog of Starburst-Galaxies as well as the expected neutrino flux was dis-

cussed in chapter 3. These sources were stacked according to the measured FIR flux at 60 μm wavelength.

- **FR-I galaxies**
  A sample of FR-I galaxies was taken from [SMAD85]. A prediction for the neutrino flux for FR-I galaxies was made in e.g. [AGHW04]. This sample was already analyzed with the AMANDA-II 4-year data set. Also here, the sources were stacked using the radio flux at 178 MHz.

- **FR-II galaxies**
  The neutrino flux of FR-II galaxies was calculated in e.g. [Bec04]. The source catalog was also acquired from [SMAD85] and has been analyzed with the 4-year data set. Like the FR-I galaxies, the FR-II galaxies were stacked according to the radio flux at 178 MHz.

- **Flat Spectrum Radio Quasars (FSRQs)**
  The analyzed sample of FSRQs was taken from recently published data measured by Fermi LAT [Abd09]. The neutrino output of FSRQs is assumed to occur in a narrow beam, calculations were done in e.g. [BB09, MSB92]. These source were stacked with the flux measured by Fermi LAT in an energy range from 1 GeV − 100 GeV.

- **Compact Steep Spectrum and Gigahertz Peaked Sources (CSS/GPS)**
  In compact sources it is believed that neutrino production takes place because the jet gets stuck in matter and nucleon-nucleon interactions take place. This source class is covered in this analysis by a sample of CSS/GPS obtained from [O'D98], it was already analyzed with the 4-year data set. The CSS/GPS sources were stacked using the radio flux measured at 1.4 GHz.

- **Blazars**
  This sample was also defined in Fermi LAT's bright sources list [Abd09], neutrino flux calculations can be found for example in [AD01] and [BB09]. The sample contains mainly BL Lac objects as well as unidentified blazars as classified in [Abd09]. These, like all source samples obtained from the Fermi bright sources list, was stacked with the flux measured by Fermi LAT in an energy range from 1 GeV − 100 GeV.

- **Pulsars**
  Pulsars are likely to appear in dense matter regions and are efficient particle accelerators [BBM05]. Pulsars as possible high energy neutrino sources have been discussed since a long time [Sat77, Bed01, Nag04, LB06], the surrounding matter serves as interaction medium for the neutrino production. As the previous source classes, this source class was stacked with the flux measured by Fermi LAT in an energy range from 1 GeV − 100 GeV. The used sample of pulsars is a sub sample of the Fermi LAT bright sources list [Abd09]. This source class as the only galactic source class was analyzed because of the measurements by Fermi LAT there was data available which was acquired very

systematically. The Fermi LAT Collaboration did not use predefined catalogs for their analysis but looked at the most significant spots on their sky map. In addition, high energy photons are a good selection criterion: If theses photons are of hadronic origin, neutrinos are bound to be produced. Recently published data on the Crab pulsar by MAGIC suggests that the photons are produced in the slot or outer gap and not in the polar cap of the pulsar [A+08b]. However, this statement needs to be confirmed by other detectors since other works [Bed01] suggest neutrino production to take place in other regions of the pulsar. Either scenario makes pulsars a source class worth being investigated.

Source classes from the 4-year analysis which were excluded in this analysis

The kev-blazars analyzed in [Gro06] were not analyzed in this analysis since these sources were out ruled by current AMANDA limits, see [A+07] for details. TeV blazars are not included in the analysis either because the study of Fermi is more systematic in terms of a catalog than TeV measurements. Fermi LAT observed the sky for 3 months while TeV instruments are pointing telescopes (IACTs) and thus bound to weather conditions. While the brightest sources are well established, the more recent discoveries are still to be completed. It is expected that more sources with the same flux density are to be detected after extended investigations with the telescopes. Another point is the variability of the sources at those wavelengths as an important factor for detection. The Fermi measurements seem to show a steady component even for variable sources giving a much better measure for the actual long-term flux which this analysis points to.

5.3 Analysis of AMANDA-II data

5.3.1 Optimization

The optimization procedure of the stacking parameters was done with a simulated E^{-2} signal spectrum and was developed in [Gro06]. The number of sources to be stacked was optimized by evaluating the signal and background for a predefined search bin of $3°$. For each source the number of signal and background events within this search bin is counted and then weighted with the measured flux which is assumed to be correlated to the neutrino flux. After processing all sources of the sample the signal is normalized to 1 for the strongest source and the signal and background is summed up resulting in a total signal ($SIG(N)$) and total background ($BG(N)$) for the sample (see equation 5.4). In order to find the best suited number of sources N the median significance $\sigma(N)$ for a signal $SIG(N)$ and a background $BG(N)$ is calculated. Poissonian statistics have to be applied here since the absolute numbers of BG and SIG are too small to apply Gaussian statistics. The probability to observe at least n events is given by

$$P(n_{\text{obs}}|SIG, BG) = \Gamma_I(n, SIG + BG), \tag{5.6}$$

where Γ_I is the incomplete Gamma function. The median of this distribution is given by n_{median} with $P(n_{\text{obs}} > n_{\text{median}}) = 0.5$. This value is obtained by inverting and resolving this equation for n_{median}. The significance of the observation of n_{median} events is given by the probability under the assumption of pure background and it is given by

$$P(n_{\text{obs}} > n_{\text{median}}|BG) = \Gamma_I(n_{\text{median}}, BG). \tag{5.7}$$

The significance in terms of a probability P is then rescaled to the corresponding number of standard deviations s of a cumulative Gaussian distribution ϕ by inverting

$$P = 1 - \phi(s) = \frac{1 - \text{erf}(s/2)}{2}. \tag{5.8}$$

Using $SIG(N)$ and $BG(N)$ in the above calculations yields $\sigma(N)$ as the significance s for N sources. The number of source to be stacked is chosen according to the maximum or the point of saturation of $\sigma(N)$ in order to maximize the significance for the analysis. The behavior of $\sigma(N)$ can be divided in three cases:

1. $\sigma(N)$ is always increasing with N: a diffuse analysis would be more sensitive for the considered hypothesis.

2. $\sigma(N)$ is always decreasing with N: the sample is dominated by the strongest source, the strongest source should be skipped in stacking and analyzed separately.

3. $\sigma(N)$ reaches a maximum or saturation at a certain N: source stacking with N sources is most sensitive for the considered hypothesis.

Here, $\sigma(N)$ was evaluated for a signal expectation of $1, 2, 3$ and 4 signal neutrinos from the strongest source. These numbers are agreeable with the point source flux limits published in [A+09b]. As search bin size in this optimization step a search bin with roughly the angular resolution of AMANDA was chosen, $3°$.

Figure 5.3 shows the significance in dependence of the number of sources for the class of the Starburst-Galaxies without the strongest source, the optimum number of sources was determined to be 12. The second parameter, the angular radius of the search bin r_{bin}, is optimized similar to the number of sources. In this optimization step the number of sources N from the first step is used. Here, the significance in dependence of r_{bin} is evaluated. Signal and background are obtained by adding the signal and background for N sources like in the step before. But here the signal is then multiplied by the value of the cumulative point spread function for the according search bin. The curve of $\sigma(r_{\text{bin}})$ shows a maximum at the optimal search bin size. This optimization step was only used for the AMANDA data and was changed for the IC-22 analysis as explained in 5.4.2.

As an example the significance in dependence of the search bin size for the

5.3. Analysis of AMANDA-II data

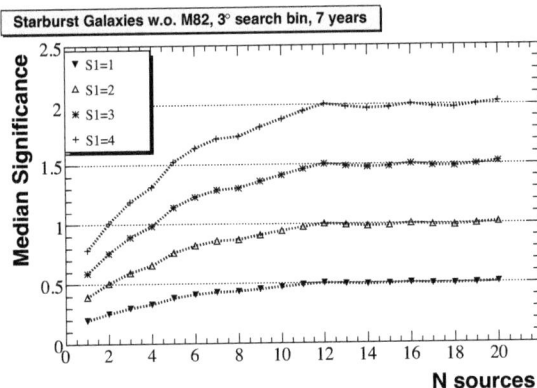

Figure 5.3: *Median significance of the sample of Starburst-Galaxies without the dominating strongest source in dependence of the number of sources. Saturation is reached for 12 sources.*

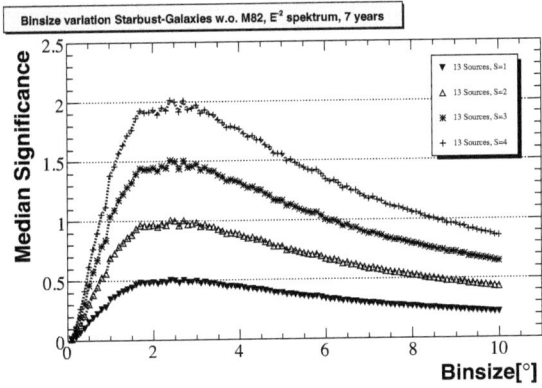

Figure 5.4: *Search bin variation for the Starburst-Galaxies. The optimal search bin is 2.4°.*

Starburst-Galaxies is shown in figure 5.4, an optimal search bin size of 2.4° was found. After having fixed the optimal r_{bin} cross-checks have been done to make sure that the previously obtained number of sources N is still the optimal number after setting the search bin size. The stacking parameters from the optimization step were tested with scrambled data sets to preserve blindness. Inside the IceCube Collaboration analyses are performed blindly, that means that all parameters of the analysis are fixed using simulated or scrambled real data. This prevents the analyses from being biased towards a certain result.

In this analysis no distance correction on the measured fluxes of the sources were done. Here, this analysis differs from the 4-year-analysis. In the 4-year analysis the fluxes were corrected to a uniform distance of $z = 0.1$. This yields a homogeneous sample where flux differences of the sources mirror the intrinsic flux properties of the sources. This correction suppresses nearby and strong sources. In this analysis no correction was applied since it is believed that nearby and strong sources in photons are also strong sources in neutrinos. This difference in both analyses leads also to different parameters for equal source classes.

5.3.2 Data

For this analysis the point source sample covering seven years of AMANDA-II data with the according Monte Carlo simulations was used. These seven years of data taking correspond to 3.8 years of detector lifetime. See [A+09b] for a detailed description of the data sample. The finalized data sample contains 6595 events. To preserve blindness only simulated and scrambled data were used. The scrambled data consist of 1000 data sets from the original data with arbitrary azimuth angles and 1000 source catalogs with randomized source positions. The significance distribution for both scrambled data sets are expected to follow a Gaussian distribution with a mean of 0 and a sigma of 1 for all source classes. The significance distribution for scrambled data sets for the Starburst-Galaxies is shown in figure 5.5(a), and the same for randomized source positions can bee seen in figure 5.5(b).

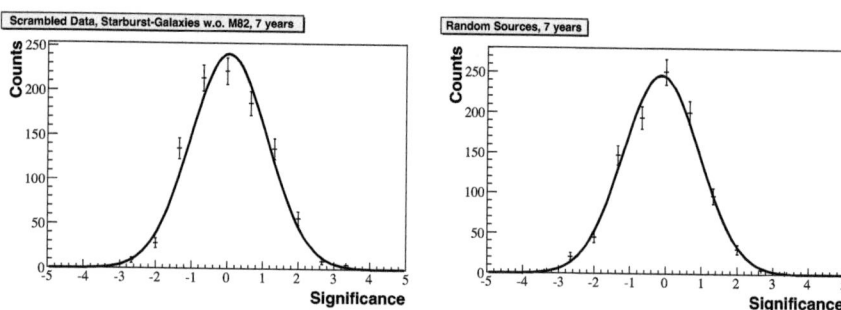

(a) Significance distribution for scrambled data for the Starburst-Galaxies

(b) Significance distribution for randomized source positions

Figure 5.5: *Significance distributions for scrambled data (left) and randomized source positions (right)*

5.3.3 The optimized parameters

As a result of the optimization process described above the optimal number of sources and the optimal search bin were obtained. These numbers are shown in table 5.1. The complete set of plots for the optimization process can be found in appendix B.1. All source classes except the

5.4. Analysis of IC-22 data

Source class	Sources	Search bin	Cat. Reference
Starbursts	13	2.4°	[BBDK09]
CSS/GPS	7	2.7°	[O'D98]
FR-I	14	2.4°	[SMAD85]
FR-II	15	2.2°	[SMAD85]
Fermi LAT blazars	12	2.4°	[Abd09]
Fermi LAT FSRQs	11	2.6°	[Abd09]
Fermi LAT pulsars	4	2.9°	[Abd09]

Table 5.1: *Results from the optimization process for seven years of AMANDA-II data.*

CSS/GPS sources had a dominating strongest source which was removed from the analysis:

- **Starburst-Galaxies** had M 82 as the dominating source.
- **FR-I Galaxies** were dominated by M 87.
- **FR-II Galaxies** were dominated by 3C 123.0.
- **Fermi LAT blazars** had 0FGL J0238.6+1636 as dominating source.
- **Fermi LAT FSRQs** were dominated by the two strongest sources, 3C 454.3 and PKS 1502+106.
- **Fermi LAT pulsars** were dominated by Geminga.

The lists of sources selected for the stacking analysis with AMANDA data can be found in appendix C.1.

5.4 Analysis of IC-22 data

The analysis described above was used on a more recent data set from the IC-22 detector. Since IC-22 has an asymmetric shape compared to the cylindrical shape of AMANDA, the analysis method had to undergo minor changes leading to different results.

5.4.1 Data

In this analysis a point source sample and according simulation files were used. The final sample contains 5114 events after the point source cuts were applied. For detailed information about the sample and the simulation files, see [A+09a]. The sample covers 275.7 days of detector lifetime. The point source analysis returned no signal for any of the analyzed sources. In figure 5.6 the effective area for different declination regions is displayed. It can be seen that IC-22 is sensitive above 100 GeV and the best acceptance is achieved for events at low declination angles i.e. near the horizon. The curve for declinations between 0° and 30° lies above all other curves.

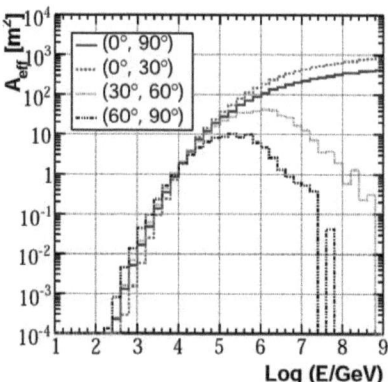

Figure 5.6: *The effective area versus the logarithm of the energy for different declination regions [GA08].*

For higher declination angles the effective area drops again at higher energies, for declination angles between 60° and 90° it reaches zero just below 100 PeV. At these high energies the neutrinos are interacting with the rocks and the Earth gets opaque for neutrinos.

5.4.2 Changes compared to the AMANDA analysis

One thing that had to be altered in this analysis was the ranking of the sources. While in the AMANDA analysis the sources were ranked according to their measured flux, the asymmetry of IC-22 had to be taken into account. This asymmetry is reflected in the zenith angle distribution of the events. In figure 5.7 there is a comparison of the AMANDA and IC-22 zenith angle distributions for simulated data. It can clearly be seen that there is an excess of events at the horizon (90°) for IC-22. This fact makes it necessary to also consider the zenith angle of the sources when ranking them. It was decided to rank the sources by a factor that takes this into account:

$$\text{Stack.Param.} = \text{measured flux} \cdot \frac{N_{\text{SIG}}}{\sqrt{N_{\text{BG}}}} \qquad (5.9)$$

with N_{SIG} being the number of signal events and N_{BG} the number of background events for each source evaluated from the simulated data. This parameter was obtained for each source and then it was normalized to 1 for the strongest source. The optimization procedure for the number of sources described above was then applied to the simulated IC-22 data with the signal weighted with the stacking parameter.

Another difference to the analysis of the AMANDA data was the optimization process for the search bin size. It turned out that the method used before delivered search bins that were

5.4. Analysis of IC-22 data

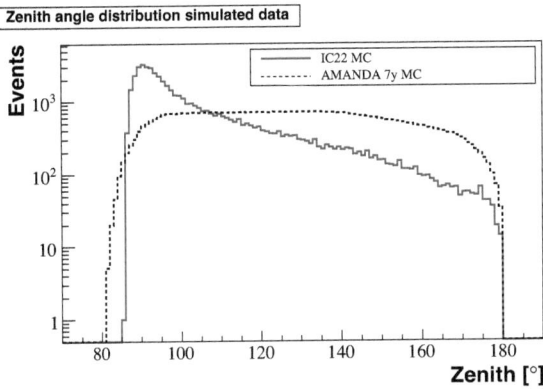

Figure 5.7: Comparison of the zenith angle distributions for simulated data between AMANDA (red) and IC-22 (black).

below IC-22's angular resolution of 1.5° [A+09a] which is not reasonable. A reason for these too small search bins could be statistical fluctuations. These fluctuations can be minimized by increasing the number of iterations over the simulated data file. Since the computing time increases linearly with the number of iterations it was unlikely to obtain more reliable results in a reasonable amount of time. Thus a different method of finding the optimal search bin size which is described in a work by Alexandreas et al. [A+93] was used. In this work the authors present a method how to calculate the optimal search bin size out of the number of background events (N_{BG}) contained in a search bin with the size of the detector's angular resolution. The optimal search bin is given by

$$r_{opt} = \left(1.58 + 0.7 e^{-0.88 \cdot N_{BG}^{0.36}}\right) \cdot \sigma. \qquad (5.10)$$

Here, σ is the angular resolution of the detector, for IC-22 this is $\sigma = 1.5°$.

5.4.3 Changes of the source selection

Due to the different ranking of the sources and the different properties of IC-22 compared to AMANDA the source selection changed. The zenith angle of the sources has a stronger influence on the ranking of the sources in the optimization process using IC-22 data. This led to slightly different source selections. The biggest change occurred for the Starburst-Galaxies and the FSRQs. Neither of the samples has a dominating strongest source in this analysis. See appendix C.2 for the source lists.

5.4.4 The optimized parameters for IC-22

In the table below (table 5.2) the number of sources and the search bin sizes obtained from the optimization process for IC-22 data are listed. It is notable that for most source classes fewer sources are selected for stacking than in the AMANDA analysis. This could be due to the fact that the sensitivity of IC-22 reaches its maximum at the horizon i.e. near a declination of $0°$. Thus sources at the near the horizon have stronger influence on the source selection while sources far from the horizon are suppressed and do not increase the sensitivity of the analysis.

Source class	Sources	Search bin
Starbursts	8	2.5°
CSS/GPS	7	2.5°
FR-I	16	2.4°
FR-II	2	2.7°
Fermi LAT blazars	9	2.5°
Fermi LAT FSRQs	2	2.7°
Fermi LAT pulsars	3	2.6°

Table 5.2: *Results from the optimization process for IC-22.*

5.5 Results

In tables 5.3 and 5.4 the results from the stacking analysis explained above with AMANDA data and IC-22 data are listed. Both analyses do not yield a significant signal and thus flux limits for each source class for a neutrino flux $\propto E^{-2}$ were set. These flux limits are median flux limits per source. The average upper limit (AUL) shown in the tables is a cumulative AUL for each source class calculated according to Feldman&Cousins [FC98] with a confidence level of 90%. This AUL was then converted into a flux limit Φ_0 for a flux proportional to E^{-2} according to

$$\Phi_0 = \text{AUL} \cdot \left(\int_{E_{\text{min}}}^{E_{\text{max}}} E^{-2} \cdot A_{\text{eff}}(E) \, dE \right)^{-1}. \tag{5.11}$$

Here, $A_{\text{eff}}(E)$ is the effective area of the detector for the whole source class depending on the energy of the incoming neutrino. The effective area is calculated using simulated data. For simulated data the ratio of incident and detected particles is known, this ratio roughly corresponds to the effective area. The energy range was $E_{\text{min}} = 50\,\text{GeV}$ to $E_{\text{max}} = 10^8\,\text{GeV}$ for AMANDA and $E_{\text{min}} = 100\,\text{GeV}$ to $E_{\text{max}} = 10^8\,\text{GeV}$ for IC-22. The flux limits are in units of $10^{-11}\,\text{TeV}^{-1}\,\text{cm}^{-2}\,\text{s}^{-1}$.

The observed under- and over-fluctuations are of statistical nature and comply with no signal

5.5. Results

being present. The last two columns of table 5.3 show the results from the 5-year analysis [A+07] and the 4-year analysis [Gro06] for source classes which were also analyzed here. The limits improved significantly compared to the results obtained with the previous AMANDA data sets, an exception is here the class of FR-I galaxies. This class shows an almost negligible better limit for the 5-year analysis. This can be explained by an under fluctuation (57 background events and 40 signal events) observed for the FR-I galaxies in the 5-year analysis. This under fluctuation is also present in the 7-year analysis since the 5-year sample is a sub sample of the 7-year sample and thus the under fluctuation does not disappear. However, this analysis did not increase the under fluctuation and so, the limit is not improved significantly.

Source class	N_{SIG}	N_{BG}	Sign. [σ]	AUL	Limit	Limit (5y)	Limit (4y)
Starbursts	76	92	1.8	16	2.0	–	–
CSS/GPS	49	51	0.2	13	1.5	7.4	7.1
FR-I	68	58	−1.4	16	1.8	1.7	2.5
FR-II	77	80	0.3	16	2.0	17.9	20.4
Fermi LAT blazars	62	56	−0.8	15	1.8	–	–
Fermi LAT FSRQs	69	66	−0.5	16	1.9	–	–
Fermi LAT pulsars	33	47	2.2	11	1.9	–	–

Table 5.3: *Results from the AMANDA analysis for an energy range of 50 GeV − 10^8 GeV. Fluxes above the limits are excluded with 90% confidence level. The last two columns show the results for the 5-year and 4-year analysis for matching source classes, the results from both analysis were calculated from integral flux limits above 10 GeV into differential flux limits.*

Source class	N_{SIG}	N_{BG}	Sign. [σ]	AUL	Limit
Starbursts	35	34	−0.2	10	1.13
CSS/GPS	28	25	−0.7	7	0.82
FR-I	61	66	0.6	20	1.03
FR-II	10	13	0.7	4	1.43
Fermi LAT blazars	39	35	−0.6	16	1.47
Fermi LAT FSRQs	11	10	−0.4	8	2.53
Fermi LAT pulsars	15	21	1.3	9	2.27

Table 5.4: *Results from the IC-22 analysis for an energy range of 100 GeV − 10^8 GeV. Fluxes above the limits are excluded with 90% confidence level.*

In the figures 5.8 and 5.9 the results of both analyses are visualized. The dashed line represents the average point source sensitivity for the northern hemisphere. The results of the analysis with AMANDA data clearly show a gain of the stacking analysis in sensitivity compared with the normal binned point source analysis. The flux limits for all source classes are well below the average sensitivity of $2.5 \cdot 10^{-11}$ TeV^{-1} cm^{-2} s^{-1}. However, the situation is different for the IC-22 analysis. In this analysis two source classes, the FSRQs and pulsars, lie above the average binned point source sensitivity of $2.0 \cdot 10^{-11}$ TeV^{-1} cm^{-2} s^{-1} [GA08]. This is due to the properties of these two source classes. Both have very few sources to stack, two FSRQs and three pulsars, which is due to the asymmetry of IC-22. For such small numbers the stacking method does not improve the results. In comparison to a new point source search method, the unbinned maximum likelihood method, the stacking analysis performed here gives improved results only for the Starburst-Galaxies, the FR-I galaxies and the CSS/GPS sources. The unbinned method is more sensitive than a binned search and is described in [A+09a], it returns an average sensitivity for the northern sky of $1.3 \cdot 10^{-11}$ TeV^{-1} cm^{-2} s^{-1}. This method searches for a maximum of the log likelihood on a search grid which is finer than the angular resolution of IC-22. The spots with a maximum in the log likelihood are then compared to source locations. The results of the IC-22 analysis performed in this thesis show that a binned stacking analysis gains sensitivity against a binned point source search except for very small samples but does not gain much against an unbinned point source search. An analysis method that combines the stacking method with an unbinned search method is planned for the new data sets.

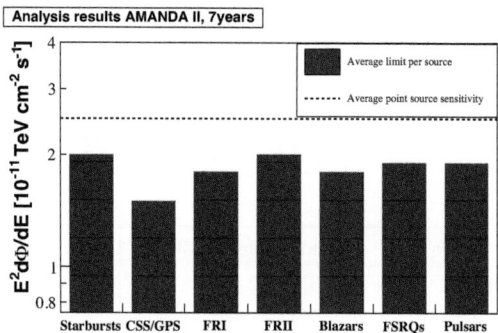

Figure 5.8: *Results of the AMANDA stacking analysis. For comparison the dashed line represents the average point source sensitivity for the northern hemisphere. All source classes gain sensitivity from the stacking method.*

5.5. Results

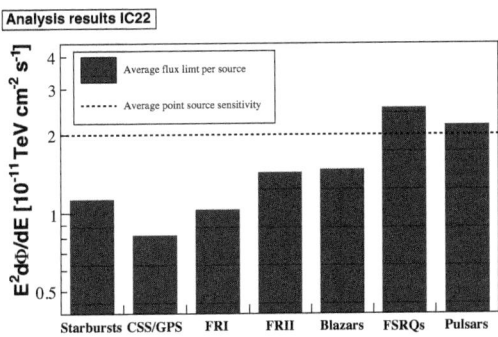

Figure 5.9: *Results of the stacking analysis of IC-22 data. The dashed line represents the average point source sensitivities from the binned IC-22 point source analysis. The stacking method gains sensitivity for all but two source classes.*

6

Conclusions & outlook

Within the scope of this thesis a source catalog of 127 local ($z < 0.03$) Starburst-Galaxies was presented and discussed in the context of neutrino astronomy. This source catalog contains measurements at different radio frequencies between 1.4 GHz and 5.10 GHz, measurements at the four FIR wavelengths measured by the IRAS satellite (12 μm, 25 μm, 60 μm and 100 μm) and measurements at X-ray energies between 0.1 keV and 8 keV. For each source the catalog contains the redshift and the luminosity distance calculated out of the redshift using the cosmological parameters $\Omega_\Lambda = 0.73$, $\Omega_{\text{Matter}} = 0.27$ and $h = 0.73$ using the on-line NASA Extragalactic Database (NED). For sources with sufficient data points in the radio a spectral index was fitted assuming a power law behavior at radio wavelengths. Sources that have sufficient data in radio, X-ray and FIR spectral indices between radio and FIR as well as between X-ray and FIR were fitted, to perform cross-checks that the selected sources indeed are starbursts.

The catalog of Starburst-Galaxies together with catalogs of AGN classes, Blazars, Flat Spectrum Radio Quasars, FR-I galaxies, FR-II galaxies, Compact Steep Spectrum / Gigahertz Peaked Sources and a catalog of Pulsars was used to perform a source stacking analysis with data of the AMANDA-II and IceCube neutrino detectors. Neither of the analyses returned a significant excess above the background of atmospheric neutrinos. Flux limits for each source class were set for a potential extraterrestrial neutrino flux following an E^{-2} spectrum. The source stacking analysis showed an improvement in sensitivity over binned point source analyses done before on the same data samples [A+09b, GA08]. Compared to an unbinned point search [A+09a] method, a gain in sensitivity exists for few source classes.

As for every piece of scientific research the experiences of the research done yield improvements for further investigations. Some ideas how to optimize the work of this thesis in the future are given in the following.

In the future, the stacking approach will be combined with the unbinned analysis method in order to improve sensitivities to point source classes even further. This method should be developed and used on future analyses with IceCube. Another analysis method worth developing is a time resolved source stacking method. This method would be sensitive to transient events i.e. Gamma Ray Bursts (GRBs). Starburst-Galaxies have an enhanced star formation rate and thus also an enhanced rate of GRBs. Restricting a time resolved stacking analysis to the sample of Starburst-Galaxies and to events occurring in a time window that matches the duration of a GRB would lead almost no background in the analysis. With such an analysis method available the analysis would also be sensitive for GRBs not detected in photons, so called choked GRBs. Choked GRBs are GRBs which happen in dense matter which is abundant in Starburst-Galaxies. For further analyses the source catalog of Starburst-Galaxies should be improved, the radio measurements presented in the catalog do often not take into account that Starburst-Galaxies are extended sources in radio astronomy. Thus wide angle measurements should be included if available. Further more it has to be checked whether the assumption of a power law behavior at radio wavelengths is justified for each source.

Although the analysis performed in this thesis and any analysis done so far revealed no extragalactic neutrino signal it is with IceCube being completed a question of time until a signal is found or current theoretical models are ruled out due to an absence of a neutrino signal. Either way, if a signal will be found or no signal will be found both would have a large impact on the neutrino astronomy and astroparticle physics.

Appendix A

The source catalog of Starburst-Galaxies

A.1 General data

Table A.1: *Position and distance data of the sample. Values obtained from NED. Coordinates epoch is J2000.0.*

Name	RA [deg]	DEC [deg]	z	D_L [Gpc]
MRK545	2.47254	25.9238	0.01523	0.05962
NGC34	2.77729	−12.1073	0.019617	0.0771
MCG-02-01-051	4.71202	−10.3768	0.027103	0.109
NGC174	9.24558	−29.4778	0.011905	0.0451
NGC232	10.6909	−23.5614	0.022172	0.0886
NGC253	11.888	−25.2882	0.0008	0.0031
IC1623	16.9466	−17.507	0.02007	0.07857
NGC520	21.1461	3.79242	0.00761	0.03022
NGC632	24.323	5.87764	0.010567	0.0396
NGC660	25.7598	13.6457	0.00283	0.01233
NGC828	32.5399	39.1904	0.01793	0.07073
NGC891	35.6392	42.3491	0.00176	0.00857
NGC958	37.6785	−2.939	0.01914	0.0765
NGC1055	40.4385	0.443167	0.00332	0.01131
Maffei2	40.4795	59.6041	$-5.7 \cdot 10^{-5}$	0.00332
NGC1068(M77)	40.6696	−0.0132806	0.00379	0.0137
UGC2238	41.5729	13.0957	0.021883	0.0883
NGC1097	41.5794	−30.2749	0.00424	0.0152
NGC1134	43.4222	13.0141	0.012142	0.0474

Table A.1: continued.

Name	R.A [deg]	DEC [deg]	z	D_L [Gpc]
NGC1365	53.4015	−36.1404	0.00546	0.01793
IC342	56.7021	68.0961	0.0001	0.0046
UGC02982	63.0935	5.54739	0.017696	0.0724
NGC1530	65.8629	75.2956	0.00821	0.03622
NGC1569	67.7044	64.8479	−0.00035	0.0046
MRK617	68.4994	−8.57888	0.01594	0.06261
NGC1672	71.4271	−59.2473	0.00444	0.01682
MRK1088	73.6598	3.26797	0.01528	0.06051
NGC1808	76.9264	−37.5131	0.00332	0.01261
NGC1797	76.937	−8.01908	0.014814	0.0616
MRK1194	77.9423	5.20061	0.01491	0.05948
NGC2146	94.6571	78.357	0.00298	0.012
NGC2276	111.81	85.7546	0.00804	0.0328
NGC2403	114.214	65.6026	0.00044	0.00247
NGC2415	114.236	35.242	0.01262	0.05341
NGC2782	138.521	40.1137	0.00848	0.03951
NGC2785	138.814	40.9175	0.008746	0.0392
NGC2798	139.346	41.9997	0.00576	0.02784
NGC2903	143.042	21.5008	0.00186	0.00826
MRK708	145.548	4.67314	0.00682	0.03116
NGC3034(M82)	148.968	69.6797	0.00068	0.00363
NGC3079	150.491	55.6797	0.00375	0.01819
NGC3147	154.224	73.4007	0.00941	0.04141
NGC3256	156.964	−43.9038	0.00935	0.03535
MRK33	158.133	54.401	0.00477	0.0221
NGC3310	159.691	53.5034	0.00331	0.01981
NGC3367	161.646	13.7509	0.010142	0.0468
NGC3448	163.663	54.3052	0.0045	0.02406
NGC3504	165.797	27.9725	0.00512	0.02707
NGC3556(M108)	167.879	55.6741	0.00233	0.01385
NGC3627(M66)	170.063	12.9915	0.00243	0.01004
NGC3628	170.071	13.5895	0.00281	0.01004
NGC3683	171.883	56.8771	0.005724	0.0259
NGC3690	172.134	58.5622	0.01041	0.04774
MRK188	176.893	55.9672	0.00803	0.0355
NGC3893	177.159	48.7108	0.00323	0.0161
NGC3994	179.404	32.2776	0.010294	0.0466

A.1. General data

Table A.1: continued.

Name	RA [deg]	DEC [deg]	z	D_L [Gpc]
NGC4030	180.099	−1.1	0.00487	0.0245
NGC4041	180.551	62.1373	0.00412	0.02278
NGC4102	181.596	52.7109	0.002823	0.0141
MRK1466	182.046	2.87828	0.00443	0.01529
MRK759	182.656	16.0329	0.00723	0.0345
NGC4194	183.539	54.5268	0.00834	0.04033
NGC4214	183.913	36.3269	0.00097	0.00367
NGC4273	184.984	5.34331	0.007932	0.0376
NGC4303(M61)	185.479	4.47365	0.005224	0.0264
NGC4414	186.613	31.2235	0.00239	0.01768
NGC4418	186.728	−0.877556	0.007268	0.0349
NGC4527	188.535	2.65381	0.005791	0.0286
NGC4536	188.613	2.18789	0.006031	0.0297
NGC4631	190.533	32.5415	0.00202	0.00773
NGC4666	191.286	−0.461885	0.005101	0.0257
NGC4793	193.67	28.9383	0.008286	0.038
NGC4826(M64)	194.182	21.6811	0.00136	0.0309
NGC4945	196.364	−49.4682	0.00187	0.00392
NGC5005	197.734	37.0592	0.00316	0.01809
NGC5020	198.166	12.5998	0.011214	0.0507
NGC5055(M63)	198.956	42.0293	0.00168	0.00796
ARP193	200.147	34.1395	0.02335	0.101
NGC5104	200.346	0.342417	0.018606	0.082
NGC5135	201.434	−29.8337	0.01372	0.05215
NGC5194(M51)	202.47	47.1952	0.00154	0.00873
NGC5218	203.043	62.7678	0.009783	0.0419
NGC5236(M83)	204.254	−29.8657	0.00172	0.00363
NGC5256	204.573	48.2769	0.027863	0.119
NGC5257	204.968	0.839583	0.022676	0.099
NGC5253	204.983	−31.6401	0.00136	0.00315
UGC8739	207.308	35.2574	0.016785	0.0728
MRK1365	208.63	15.0441	0.01846	0.0806
NGC5430	210.191	59.3283	0.009877	0.0423
NGC5427	210.859	−6.03081	0.008733	0.0399
NGC5678	218.023	57.9214	0.00641	0.03202
NGC5676	218.195	49.4579	0.007052	0.0308
NGC5713	220.048	−0.289222	0.00658	0.02674

Table A.1: continued.

Name	RA [deg]	DEC [deg]	z	D_L [Gpc]
NGC5775	223.49	3.54446	0.00561	0.02634
NGC5900	228.772	42.2094	0.008376	0.0361
NGC5936	232.504	12.9893	0.013356	0.0575
ARP220	233.738	23.5032	0.01813	0.0799
NGC5962	234.132	16.6079	0.006528	0.0288
NGC5990	236.568	2.41547	0.012806	0.055
NGC6181	248.087	19.8266	0.007922	0.0334
NGC6217	248.163	78.1982	0.00454	0.02349
NGC6240	253.245	2.40094	0.02448	0.10336
NGC6286	254.631	58.9363	0.018349	0.0761
IRAS18293-3413	278.171	−34.191	0.01818	0.07776
NGC6701	280.802	60.6533	0.01323	0.05664
NGC6764	287.068	50.9332	0.008059	0.03131
NGC6946	308.718	60.1539	0.00016	0.00532
NGC7130	327.081	−34.9513	0.01615	0.06599
IC5179	334.038	−36.8437	0.01141	0.0467
NGC7331	339.267	34.4156	0.00272	0.01471
NGC7469	345.815	8.874	0.01632	0.06523
NGC7479	346.236	12.3229	0.00794	0.03236
NGC7496	347.447	−43.4279	0.0055	0.02234
NGC7541	348.683	4.53436	0.008969	0.032
IC5298	349.003	25.5567	0.027422	0.11
NGC7552	349.045	−42.5848	0.00536	0.02144
NGC7591	349.568	6.58581	0.016531	0.0636
NGC7592	349.592	−4.41694	0.024444	0.0972
MRK319	349.66	25.2329	0.027012	0.108
NGC7673	351.921	23.5889	0.01137	0.0422
NGC7678	352.116	22.4212	0.011639	0.0433
MRK534	352.194	3.51142	0.01714	0.0677
NGC7679	352.194	3.51142	0.017139	0.0662
NGC7714	354.059	2.15516	0.00933	0.0386
NGC7771	357.854	20.1118	0.01427	0.05711
NGC7793	359.458	−32.591	0.00076	0.0031
MRK332	359.856	20.7499	0.00802	0.0283

A.2 Radio data

Table A.2: Radio measurements of the sample. All fluxes in [mJy].
References:

1: [RRA91], 2: [CCB02], 3: [CCBD83], 4: [JWE+98], 5: [GWBE94], 6: [JGDB96], 7: [WGH+96], 8: [DC78], 9: [GAO+06], 10: [Sra75], 11: [BBB+04], 12: [RSTT07], 13: [GWBE95], 15: [VTW04], 16: [AR90], 17: [KWPN81], 18: [WGBE94], 19: [RR92], 20: [RRD95], 21: [SHC+04], 22: [NFW05], 24: [DW77], 25: [LBSB05], 26: [Con83], 27: [IYH05], 28: [ST76]

Name	$S_{1.40\,GHz}$	$S_{2.38\,GHz}$	$S_{2.69\,GHz}$	$S_{2.70\,GHz}$	$S_{4.85\,GHz}$	$S_{5.00\,GHz}$	$S_{5.01\,GHz}$	References
MRK545	73.5	47	–	–	33	36	–	1, 2, 8, 14
NGC34	67.5	–	–	–	–	–	–	4
MCG-02-01-051	43.2	–	–	–	–	–	–	4
NGC174	45.7	–	–	–	–	–	–	4
NGC232	60.6	–	–	–	56	–	–	4
NGC253	6000	–	–	3520	2433	–	2580	5, 16, 17
IC1623	249.2	–	–	–	96	–	–	4, 5
NGC520	176	110	–	–	87	–	–	1, 2, 8
NGC632	23	15	–	–	–	–	–	2, 8
NGC660	373	255	–	–	187	–	–	1, 4, 8
NGC828	108	–	–	–	47	–	–	6
NGC891	701	–	–	–	342	–	–	2
NGC958	71.9	–	–	–	–	–	–	4
NGC1055	200.9	129	–	150	63	–	–	1, 4, 8, 16
Maffei2	1015	–	–	–	375	–	–	1, 19
NGC1068(M77)	4850	–	–	3050	2039	1890	1342.4	2, 9, 13, 14, 17
UGC2238	72.2	–	–	–	–	–	–	1, 2
NGC1097	415	–	–	250	126	–	150	6, 7, 16
NGC1134	89.1	57	–	–	32	–	–	1, 2, 8
NGC1365	530	–	–	350	230	180	–	4, 7, 16

Table A.2: continued.

Name	$S_{1.40\,\text{GHz}}$	$S_{2.38\,\text{GHz}}$	$S_{2.69\,\text{GHz}}$	$S_{2.70\,\text{GHz}}$	$S_{4.85\,\text{GHz}}$	$S_{5.00\,\text{GHz}}$	$S_{5.01\,\text{GHz}}$	References
IC342	2250	–	–	–	277	–	–	1,6
UGC02982	91.3	–	–	–	–	–	–	2
NGC1530	80.7	–	–	–	27	–	–	1,6
NGC1569	396	–	–	–	198	–	155	1,10,19
MRK617	138	–	–	–	63	–	–	4,13
NGC1672	450	–	–	210	114	–	100	16,18
MRK1088	45.7	31	–	–	–	–	–	2,8
NGC1808	497	–	–	350	229	–	220	6,16,18
NGC1797	29.1	–	–	–	–	–	–	4
MRK1194	42.2	27	–	–	–	–	–	2,8
NGC2146	1087	–	–	–	–	–	–	6
NGC2276	283	–	–	–	–	–	–	6
NGC2403	387	–	–	–	169	–	–	19
NGC2415	66.4	53	–	–	41	–	30	1,2,8,10
NGC2782	107.5	–	–	–	47	–	60	1,10,20
NGC2785	67.6	–	–	–	–	–	–	2
NGC2798	82.0	–	–	–	37	–	53	1,2,10
NGC2903	444	200	–	–	118	–	–	1,2,8,13
MRK708	32.6	21	–	–	–	–	–	2,8
NGC3034(M82)	7286.8	–	5650	–	3918	–	3912	1,17,20
NGC3079	820.7	–	–	–	321	–	–	1,21
NGC3147	89.9	–	–	–	44	–	8.1	1,2,23
NGC3256	642	–	–	–	319	240	250	5,6,14,16
MRK33	24.6	–	–	–	–	–	–	11

Table A.2: continued.

Name	$S_{1.40\,GHz}$	$S_{2.38\,GHz}$	$S_{2.69\,GHz}$	$S_{2.70\,GHz}$	$S_{4.85\,GHz}$	$S_{5.00\,GHz}$	$S_{5.01\,GHz}$	References
NGC3310	417	–	–	–	152	–	–	1, 19
NGC3367	118	71	–	130	36	–	35	1, 2, 8, 10, 16
NGC3448	51.3	–	–	–	–	–	39	1, 10
NGC3504	274	–	–	–	117	–	–	1, 2, 8
NGC3556(M108)	245	–	–	–	76	–	–	1, 19
NGC3627(M66)	458	209	–	–	141	–	–	1, 2, 8
NGC3628	470.2	313	–	–	276	200	224	1, 8, 10, 14, 21
NGC3683	127	–	–	–	–	–	–	2
NGC3690	658	–	–	–	–	–	362	10, 23
MRK188	30.7	–	–	–	–	–	25.0	2, 10
NGC3893	139	–	–	–	39	–	–	1, 2
NGC3994	70.8	50	–	–	52	–	–	1, 2, 8
NGC4030	147	–	–	90	–	56	–	14, 16, 19
NGC4041	103	–	–	–	48	–	–	1, 2
NGC4102	273	–	–	–	70	–	105	1, 10
MRK1466	20.0	15	–	–	–	–	–	2, 8
MRK759	31.9	16	–	–	13	–	–	2, 8, 15
NGC4194	122	–	–	–	39	–	–	1, 19
NGC4214	38.3	–	–	–	30	–	–	2
NGC4273	78.5	65	–	–	37	–	–	2, 8, 15
NGC4303(M61)	444	195	–	–	102	120	–	2, 8, 14, 15
NGC4414	227	138	–	–	75	–	–	1, 8, 19
NGC4418	38.5	–	–	–	–	–	–	23
NGC4527	187.9	129	–	–	72	–	–	2, 8, 15

Table A.2: continued.

Name	$S_{1.40\,GHz}$	$S_{2.38\,GHz}$	$S_{2.69\,GHz}$	$S_{2.70\,GHz}$	$S_{4.85\,GHz}$	$S_{5.00\,GHz}$	$S_{5.01\,GHz}$	References
NGC4536	204.9	136	–	–	114	110	–	2, 8, 13
NGC4631	1122	340	–	–	438	–	–	1, 8, 19
NGC4666	434	–	–	–	161	–	–	2, 13
NGC4793	113	72	–	–	46	–	–	1, 2, 8
NGC4826(M64)	103	67	–	–	54	–	–	1, 2, 8
NGC4945	6600	–	–	5000	3055	–	2840	16, 18
NGC5005	194	–	–	–	62	–	–	1, 19
NGC5020	30.1	22	–	–	–	–	–	2, 8
NGC5055(M63)	349	–	–	–	124	–	–	1, 2
ARP193	104	–	–	–	53	–	–	1, 2
NGC5104	39.9	38	–	–	–	–	–	2, 8
NGC5135	194	–	–	–	107	–	–	6, 7
NGC5194(M51)	1310	–	–	–	436	–	360	1, 10, 19
NGC5218	30.4	–	–	–	–	–	–	2
NGC5236(M83)	2445	–	–	–	648	–	–	6, 7, 16
NGC5256	159	–	–	–	47	–	–	1, 19
NGC5257	48.7	48	–	–	–	–	–	2, 8
NGC5253	83.8	–	–	–	90	75	–	6, 7, 14
UGC8739	93.6	–	–	–	37	–	–	1, 2
MRK1365	23.0	14	–	–	–	–	–	2, 8
NGC5430	65.9	–	–	–	29	–	40	1, 2, 10
NGC5427	63	–	–	–	–	–	–	26
NGC5678	110	–	–	–	68	–	–	2
NGC5676	116	–	–	–	38	–	33	2

Table A.2: continued.

Name	$S_{1.40\,GHz}$	$S_{2.38\,GHz}$	$S_{2.69\,GHz}$	$S_{2.70\,GHz}$	$S_{4.85\,GHz}$	$S_{5.00\,GHz}$	$S_{5.01\,GHz}$	References
NGC5713	222	–	–	–	93	73	–	1, 2, 14
NGC5775	221	138	–	–	67	–	–	1, 8, 19
NGC5900	60.4	–	–	–	–	17	–	2, 10
NGC5936	139	81	–	–	48	–	59	1, 2, 8, 10
ARP220	324	–	260	–	208	–	–	1, 2, 3
NGC5962	82.3	56	–	–	36	–	–	1, 2, 8
NGC5990	63.9	39	–	–	–	–	–	2, 8
NGC6181	95.6	60	–	–	56	–	–	1, 2, 8
NGC6217	79.9	–	–	–	–	–	–	2
NGC6240	653	–	–	–	179	–	170	13, 16, 19
NGC6286	157	–	–	–	53	–	–	1, 2
IRAS18293-3413	223	–	–	–	144	–	–	6, 7
NGC6701	92.2	–	–	–	22	–	–	1, 6
NGC6764	115.0	–	–	–	34	–	47	1, 2, 10
NGC6946	1395	–	–	–	531	–	–	1, 19
NGC7130	183	–	–	–	–	–	–	6
IC5179	165	–	–	–	79	–	–	6, 7
NGC7331	373	187	–	–	80	–	–	1, 2, 8
NGC7469	255	132	–	–	95	–	–	8, 13, 19
NGC7479	107	58	–	–	41	–	–	1, 2, 8
NGC7496	36.3	–	–	–	–	–	–	6
NGC7541	162	101	–	–	56	–	–	1, 2, 8
IC5298	24.2	–	–	–	–	–	–	23
NGC7552	276	–	–	–	139	–	–	6, 18

A.2. Radio data

Table A.2: continued.

Name	$S_{1.40\,GHz}$	$S_{2.38\,GHz}$	$S_{2.69\,GHz}$	$S_{2.70\,GHz}$	$S_{4.85\,GHz}$	$S_{5.00\,GHz}$	$S_{5.01\,GHz}$	References
NGC7591	52.1	39	–	–	–	–	–	2,8
NGC7592	75	–	–	–	–	–	–	27
MRK319	31.6	21	–	–	–	–	–	2,8
NGC7673	43.4	30	–	–	–	–	–	2,8
NGC7678	49.5	36	–	–	–	–	–	2,8
MRK534	55.8	33	–	–	45	–	–	2,8,13
NGC7679	55.8	33	–	–	45	–	–	2,8,13
NGC7714	65.8	–	–	–	39	–	–	1,2
NGC7771	229	97	–	–	57	–	–	1,8,19
NGC7793	103	–	–	–	–	–	–	6
MRK332	36.5	27	–	–	–	–	–	2,8

A.3 Far Infrared data

Table A.3: IRAS measurements of the sample. All fluxes in [Jy].
References:
1: [SMK+03], 2: [MG+90], 3: [LVS+07], 4: [SSM04], 5: [Kna94], 6: [BNH+88], 7: [SBNS89]

Name	$S_{12\,\mu m}$	$S_{25\,\mu m}$	$S_{60\,\mu m}$	$S_{100\,\mu m}$	References
MRK545	0.523	1.082	9.196	15.34	2
NGC34	0.35	2.39	17.05	16.86	1
MCG-02-01-051	0.24	1.19	7.35	10.22	4
NGC174	0.41	1.27	11.36	19.77	1
NGC232	0.36	1.28	10.05	17.14	1
NGC253	41.04	154.67	967.81	1288.15	1
IC1623	1.03	3.65	22.93	31.55	1
NGC520	0.9	3.22	31.52	47.37	2
NGC632	0.37	0.88	4.89	7.32	5
NGC660	3.05	7.3	65.52	114.74	1
NGC828	0.72	1.07	11.46	25.33	1
NGC891	5.27	7	66.46	172.23	1
NGC958	0.62	0.94	5.85	15.08	1
NGC1055	2.24	2.84	23.37	65.26	1
Maffei2	3.624	9.238	135	225	6
NGC1068(M77)	39.84	87.57	196.37	257.37	1
UGC2238	0.36	0.65	8.17	15.67	1
NGC1097	2.96	7.3	53.35	104.79	1
NGC1134	0.55	0.92	9.09	19.43	1
NGC1365	5.12	14.28	64.31	165.67	1
IC342	14.92	34.48	180.8	391.66	1
UGC02982	0.57	0.83	8.39	16.82	1
NGC1530	0.72	1.23	9.88	25.88	1
NGC1569	1.24	9.03	54.36	55.29	1
MRK617	0.441	7.286	32.31	32.69	1
NGC1672	2.47	5.25	41.21	77.92	1
MRK1088	0.2659	0.835	6.605	10.77	1
NGC1808	5.4	17	105.55	141.76	1
NGC1797	0.33	1.35	9.56	12.76	1
MRK1194	0.283	0.7071	6.688	11.5	2
NGC2146	6.83	18.81	146.69	194.05	1
NGC2276	1.07	1.63	14.29	28.97	1

A.3. Far Infrared data

Table A.3: continued.

Name	$S_{12\mu m}$	$S_{25\mu m}$	$S_{60\mu m}$	$S_{100\mu m}$	References
NGC2403	2.82	3.57	41.47	99.13	1
NGC2415	0.61	1.19	8.75	13.58	1
NGC2782	0.64	1.51	9.17	13.76	1
NGC2785	0.49	1.09	8.4	15.79	1
NGC2798	0.76	3.21	20.6	29.69	2
NGC2903	5.29	8.64	60.54	130.43	1
MRK708	0.46	0.8	5.36	8.24	1
NGC3034(M82)	79.43	332.63	1480.42	1373.69	1
NGC3079	2.54	3.61	50.67	104.69	1
NGC3147	1.95	1.03	8.17	29.61	1
NGC3256	3.57	15.69	102.63	114.31	1
MRK33	0.21	1.05	4.77	5.99	5
NGC3310	1.54	5.32	34.56	44.19	1
NGC3367	0.51	1.98	6.44	13.48	1
NGC3448	0.22	0.64	6.64	11.17	1
NGC3504	1.11	4.03	21.43	34.05	1
NGC3556(M108)	2.29	4.19	32.55	76.9	1
NGC3627(M66)	4.82	8.55	66.31	136.56	1
NGC3628	3.13	4.85	54.8	105.76	1
NGC3683	1.16	1.48	13.87	29.3	1
NGC3690	3.97	24.51	113.05	111.42	2
MRK188	0.3621	0.4515	4.576	11.52	2
NGC3893	1.45	1.65	15.57	36.8	1
NGC3994	0.32	0.46	4.98	10.31	4
NGC4030	1.35	2.3	18.49	50.92	1
NGC4041	1.13	1.56	14.15	31.74	2
NGC4102	1.77	6.83	46.85	70.29	1
MRK1466	0.325	1.236	6.265	10.52	2
MRK759	0.2995	0.537	4.116	8.727	2
NGC4194	0.99	4.51	23.2	25.16	1
NGC4214	0.58	2.46	17.57	29.08	2
NGC4273	0.77	1.65	9.38	21.76	1
NGC4303(M61)	3.28	4.9	37.27	78.74	1
NGC4414	2.78	3.61	29.55	70.69	1
NGC4418	0.99	9.67	43.89	31.97	1
NGC4527	2.65	3.55	31.4	65.68	1
NGC4536	1.55	4.04	30.26	44.51	1

Table A.3: continued.

Name	$S_{12\,\mu m}$	$S_{25\,\mu m}$	$S_{60\,\mu m}$	$S_{100\,\mu m}$	References
NGC4631	5.16	8.97	85.4	160.08	1
NGC4666	3.34	3.89	37.11	85.95	1
NGC4793	1.08	1.57	12.42	28.11	1
NGC4826(M64)	2.36	2.86	36.7	81.65	1
NGC4945	27.47	42.34	625.46	1329.7	1
NGC5005	1.65	2.26	22.18	63.4	1
NGC5020	0.36	0.72	5.58	11.7	1
NGC5055(M63)	5.35	6.36	40	139.82	1
ARP193	0.25	1.42	17.04	24.38	1
NGC5104	0.39	0.74	6.78	13.37	1
NGC5135	0.63	2.38	16.86	30.97	1
NGC5194(M51)	7.21	9.56	97.42	221.21	1
NGC5218	0.37	0.94	7.01	13.54	1
NGC5236(M83)	21.46	43.57	265.84	524.09	1
NGC5256	0.32	1.07	7.25	10.11	1
NGC5257	0.52	1.18	8.1	13.63	4
NGC5253	2.612	12.07	29.84	30.08	2
UGC8739	0.35	0.42	5.79	15.89	1
MRK1365	0.1562	0.6445	4.203	6.113	2
NGC5430	0.5	1.94	10.1	20.34	1
NGC5427	1.29	1.48	10.24	25.29	1
NGC5678	0.94	1.2	9.67	25.66	1
NGC5676	1.13	1.7	12.04	29.91	1
NGC5713	1.47	2.84	22.1	37.28	1
NGC5775	1.83	2.47	23.59	55.64	1
NGC5900	0.4	0.7	7.51	16.95	1
NGC5936	0.48	1.47	8.73	17.66	1
ARP220	0.61	8	104.09	115.29	1
NGC5962	0.73	1.04	8.93	21.82	1
NGC5990	0.6	1.6	9.59	17.14	1
NGC6181	0.63	1.41	8.94	20.83	1
NGC6217	0.74	2.03	11.35	20.62	1
NGC6240	0.59	3.55	22.94	26.49	1
NGC6286	0.47	0.62	9.24	23.11	1
IRAS18293-3413	1.14	3.98	35.71	53.38	1
NGC6701	0.55	1.32	10.05	20.05	1
NGC6764	0.54	1.33	6.62	12.44	1

Table A.3: continued.

Name	$S_{12\,\mu m}$	$S_{25\,\mu m}$	$S_{60\,\mu m}$	$S_{100\,\mu m}$	References
NGC6946	12.11	20.7	129.78	290.69	1
NGC7130	0.58	2.16	16.71	25.89	1
IC5179	1.18	2.4	19.39	37.29	1
NGC7331	3.94	5.92	45	110.16	1
NGC7469	1.59	5.96	27.33	35.16	1
NGC7479	1.37	3.86	14.93	26.73	1
NGC7496	0.58	1.93	10.14	16.57	1
NGC7541	1.52	2.09	20.08	41.87	1
IC5298	0.34	1.95	9.06	11.99	1
NGC7552	3.76	11.92	77.37	102.92	1
NGC7591	0.28	1.27	7.87	14.87	1
NGC7592	0.26	0.97	8.05	10.58	1
MRK319	0.2211	0.5418	4.266	7.062	2
NGC7673	0.1329	0.5165	4.98	6.893	1, 2
NGC7678	0.63	1.16	6.98	14.84	1
MRK534	0.5	1.12	7.4	10.71	1
NGC7679	0.5	1.12	7.58	10.71	1
NGC7714	0.47	2.88	11.16	12.26	1
NGC7771	0.99	2.17	19.67	40.12	1
NGC7793	1.32	1.67	18.14	54.07	1
MRK332	0.3598	0.6212	4.871	9.493	2

A.4 X-ray data

A.4. X-ray data

Table A.4: X-Ray measurements of the sample. All fluxes in [nJy].
References:
1: [RLR96], 2: [FKT92], 3: [WGA00], 4: [TTW+05], 5: [BSB94], 6: [OWB05], 7: [TPR+06], 8: [TWV+05], 9: [S+07], 10: [GMP05]

Name	EO IPC 0.1-4 keV	EINSTEIN 0.2-4.0 keV	Chandra 0.1-2.4 keV	ROSAT 0.2-2.0 keV	ROSAT 0.1-2.4 keV	Chandra 0.3-8 keV	XMM 0.3-2 keV	References
NGC34	–	–	–	–	–	–	23	9
NGC520	–	–	–	33.4	–	–	–	3
NGC660	–	–	–	45.1	–	–	–	3
NGC891	–	–	–	–	–	–	69.4	7
NGC1068(M77)	–	3940	–	–	10900	–	–	2,5
NGC1097	–	578	–	–	–	–	–	2
NGC1365	–	326	–	–	–	–	–	2
IC342	–	939	–	–	–	–	–	2
NGC1569	–	368	–	–	–	–	–	2
MRK617	–	124	–	–	–	–	–	2
MRK1088	–	–	–	42.3	–	–	–	8
NGC1808	–	–	–	–	446	–	–	5
NGC2403	–	364	–	–	–	–	–	2
NGC2415	–	–	–	63.8	–	–	–	3
NGC2903	–	273	–	46.4	–	–	–	2,3
NGC3034(M82)	–	4490	–	–	–	–	–	2
NGC3079	–	113	–	–	–	–	–	2
NGC3256	–	–	–	–	1060	–	–	5
NGC3310	–	208	–	–	–	–	–	2
NGC3367	102	–	–	–	–	–	–	1
NGC3448	–	72.6	–	–	–	–	–	2

Table A.4: continued.

Name	EO IPC 0.1-4 keV	EINSTEIN 0.2-4.0 keV	Chandra 0.1-2.4 keV	ROSAT 0.2-2.0 keV	ROSAT 0.1-2.4 keV	Chandra 0.3-8 keV	XMM 0.3-2 keV	References
NGC3504	–	76.8	–	–	–	–	–	2
NGC3627(M66)	–	–	–	–	603	–	–	5
NGC3690	–	91.5	–	–	–	–	–	2
NGC4102	–	–	–	–	–	–	–	9
NGC4214	–	–	–	–	–	75.6	45.1	6
NGC4273	–	–	–	67.1	–	–	–	3
NGC4303(M61)	–	180	–	–	233	–	–	2,5
NGC4536	–	112	–	–	–	–	–	2
NGC4631	–	263	–	–	–	–	–	2
NGC4826(M64)	–	150	–	53.3	–	–	–	3
NGC4945	–	–	–	–	794	–	–	5
NGC5135	–	63.9	–	176	–	–	–	3
NGC5236(M83)	–	972	–	–	954	–	–	2,5
NGC5256	–	–	–	–	–	–	9.93	10
NGC5253	–	41.4	–	38.5	–	35.8	–	2,3,6
NGC5775	–	–	–	–	–	–	15.8	7
ARP220	45.69	–	–	–	–	–	–	1
NGC6240	–	–	–	–	914	–	–	5
NGC6946	–	611	–	–	–	–	–	2
NGC7331	–	16.8	–	76.7	–	–	–	2,3
NGC7469	–	12000	–	–	–	–	–	2
NGC7552	–	167	–	–	–	–	–	2
MRK534	–	204	–	–	–	–	–	3

A.4. X-ray data

Table A.4: continued.

Name	EO IPC 0.1-4 keV	EINSTEIN 0.2-4.0 keV	Chandra 0.1-2.4 keV	ROSAT 0.2-2.0 keV	ROSAT 0.1-2.4 keV	Chandra 0.3-8 keV	XMM 0.3-2 keV	References
NGC7679	–	204	–	–	–	–	–	2
NGC7714	–	53.8	–	–	–	–	–	2
NGC7771	–	138	–	–	–	–	–	2
NGC7793	–	130	–	–	–	–	–	2

Appendix B

Supplementary plots

B.1 Optimization plots of the AMANDA analysis

In the following the plots optained during the optimization process are shown: Relative strength of the sources ,significance versus number of sources, Significance versus search bin size and significance distribution for scrambled datasets.

Starburst-Galaxies

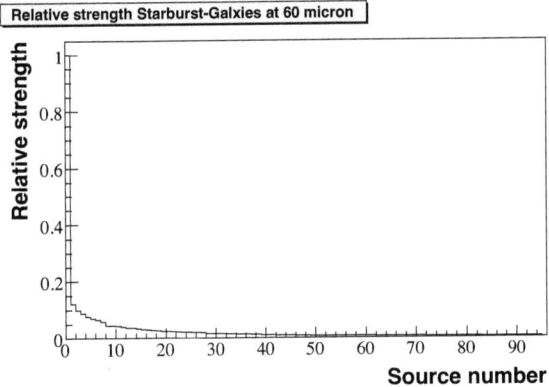

Figure B.1: The relative strength of the sample of Starburst-Galaxies at a wavelength of 60 μm. The sources are ordered according to their flux, the strongest source, M82, has the value 1.

The figures B.2 and B.3 show the optimization of the number of sources. Figure B.2 shows that the sample is dominated by the strongest source in the sample (M82). Therefore, for the further analysis, this source was skipped resulting in figure B.3. The optimal number of sources to stack is 13. Figure B.4 shows the next step of the optimization. The curves show the

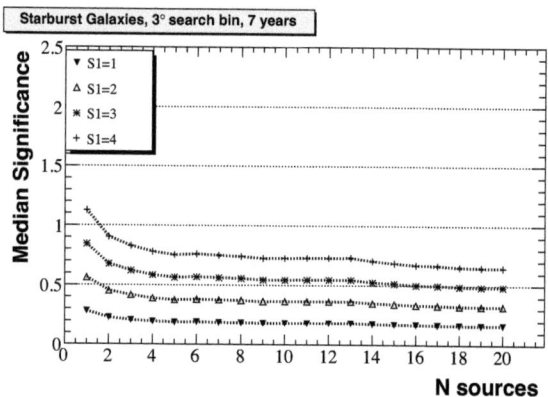

Figure B.2: Median significance of the sample of Starburst-Galaxies in dependence of the number of sources for $1, 2, 3$ or 4 signal neutrinos from the strongest source. The significance is monotonously decreasing, the sample is dominated by the strongest source.

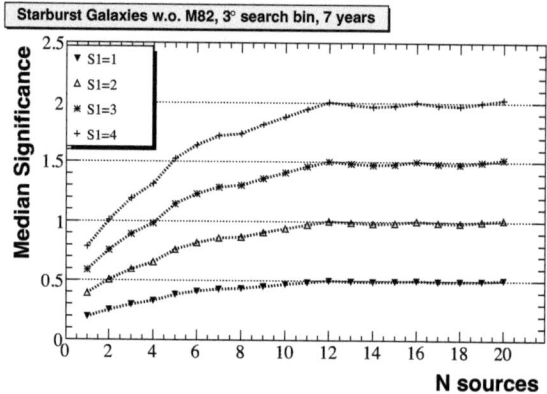

Figure B.3: The same figure as figure B.2 but without the strongest source. The significance reaches a maximum at 13 sources.

median significance of the sample in dependence of the search bin size for 13 sources stacked. A maximum is reached for a search bin size of 2.4°. After fixing the stacking parameters the sample was analyzed using scrambled data. As expected the significance of the 1000 analyzed data sets follows a Gaussian distribution centered around 0.

B.1. Optimization plots of the AMANDA analysis

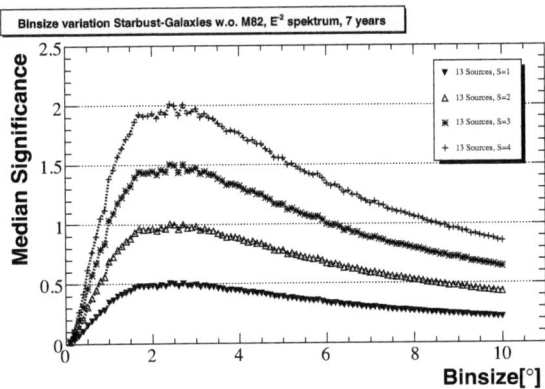

Figure B.4: Median significance for different search bins. The optimal search bin size is 2.4°.

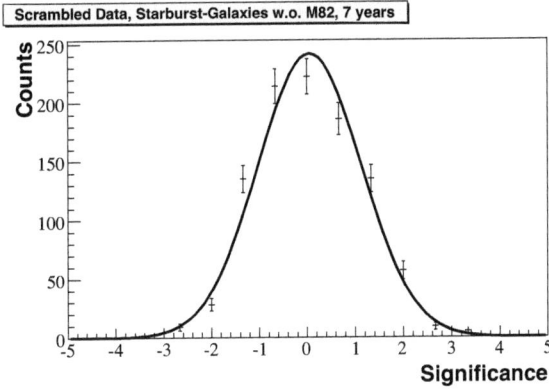

Figure B.5: Distribution of the median significance for 1000 scrambled data sets. It follows a Gaussian distribution with a mean of 0.

CSS/GPS sources

These compact sources were stacked according to their radio flux measured at 1.4 GHz. Radio emissions at this frequency is synchrotron radiation of primary electrons. This serves as a proof that charged particles are accelerated in the source. Thus it is likely that also protons are accelerated. Since this source class contains compact sources these accelerated protons can undergo proton proton interactions in which neutrinos are produced. The relative strength of the sources is shown in figure B.6. The figures B.7, B.8 and B.9 show the optimization process resulting in 7 sources to stack with a search bin of 2.7°.

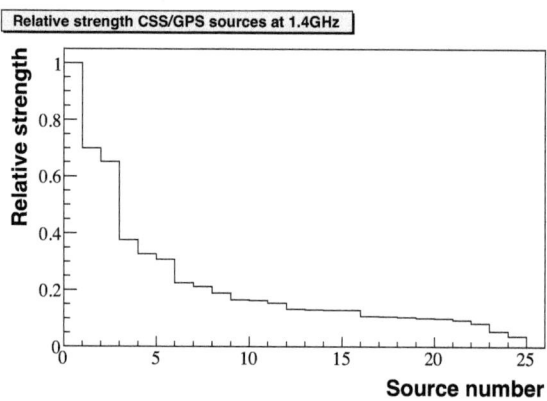

Figure B.6: The relative strength of the CSS/GPS sources at 1.4 GHz. The sources are ordered according to their flux, the strongest source has the value 1.

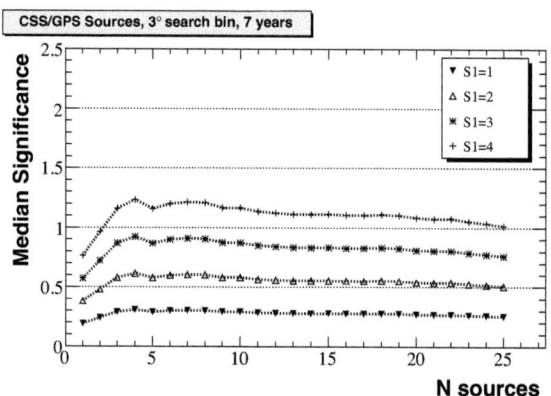

Figure B.7: Median significance of the sample of CSS/GPS sources in dependence of the number of sources. Maximum is reached for 7 sources.

B.1. Optimization plots of the AMANDA analysis

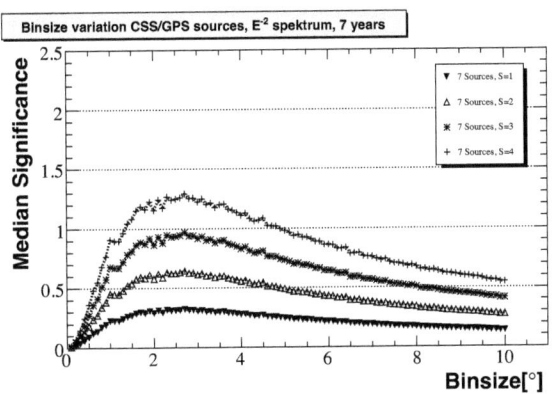

Figure B.8: Search bin variation for CSS/GPS sources. The optimal search bin is 2.4°.

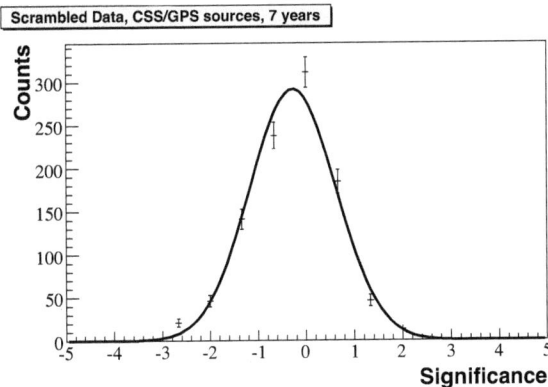

Figure B.9: Distribution of the median significance for 1000 scrambled data sets. It follows a Gaussian distribution with a mean of 0.

FR-I and FR-II galaxies

These sources were reanalyzed and the samples were taken from [SMAD85] and stacked with the radio flux at 178 MHz. Both samples were dominated by the strongest sources, M 87 and 3C123 respectively. The optimization process yields 14 sources to stack with a search bin of 2.4° for the FR-I galaxies and 15 sources and 2.2° for the FR-II galaxies. The relative strengths in the samples are in figure B.10 and B.11. Figures B.12, B.13, B.14 and B.15 show the optimization of the number of sources, the size of the search bin and the results for scrambled data for the FR-I galaxies. The according plots for the FR-II galaxies are in the figures B.16, B.17, B.18 and B.19.

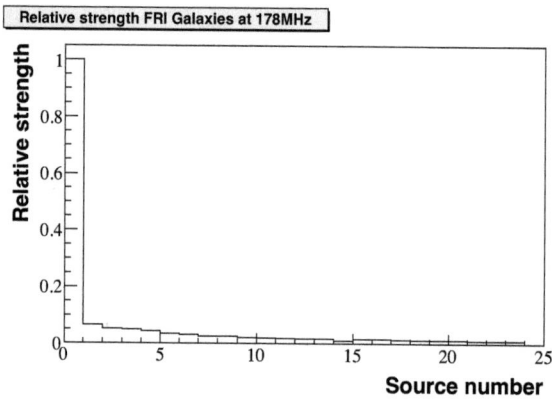

Figure B.10: The relative strength of the FR-I galaxies at 178 MHz.

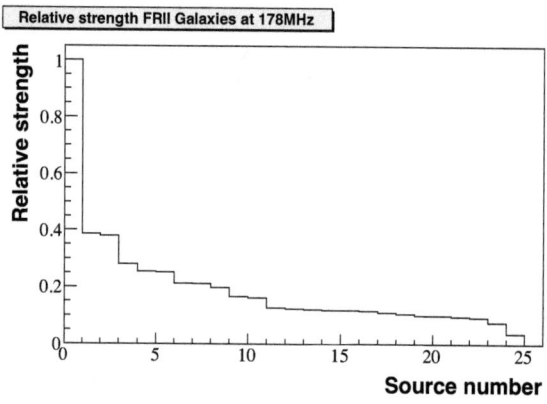

Figure B.11: The relative strength of the FR-II at 178 MHz.

B.1. Optimization plots of the AMANDA analysis

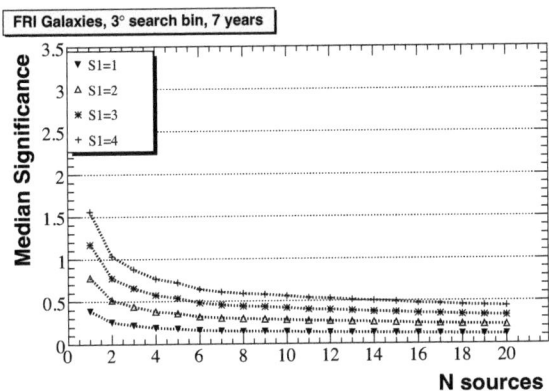

Figure B.12: Median significance of the sample of FR-I sources in dependence of the number of sources. The sample is dominated by M87.

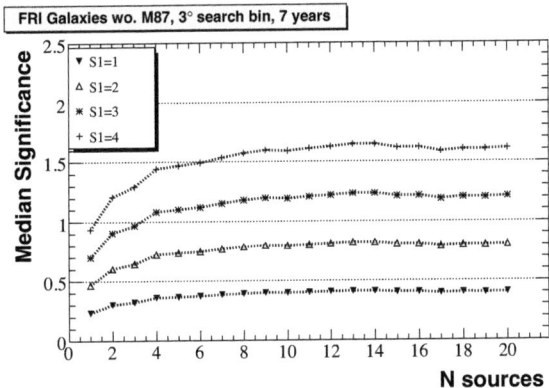

Figure B.13: Median significance of the sample of FR-I sources without M87 in dependence of the number of sources. The optimal number of sources is 14.

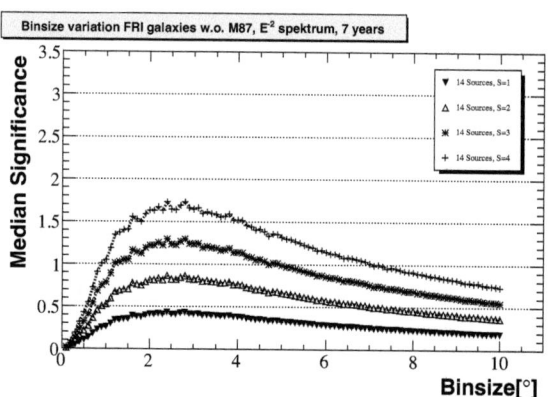

Figure B.14: Search bin variation for FR-I sources. The optimal search bin is 2.4°.

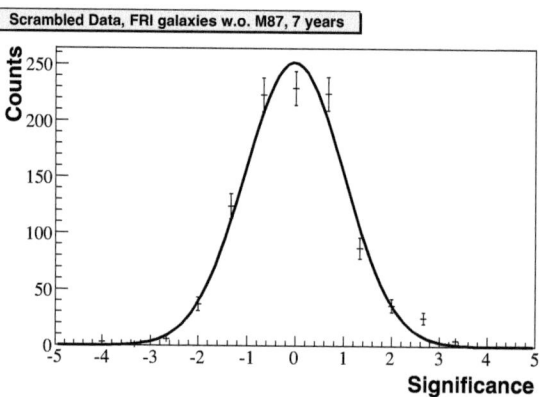

Figure B.15: Analysis of the sample with 1000 scrambled data sets. The significance follows a Gaussian distribution with a mean of 0.

B.1. Optimization plots of the AMANDA analysis

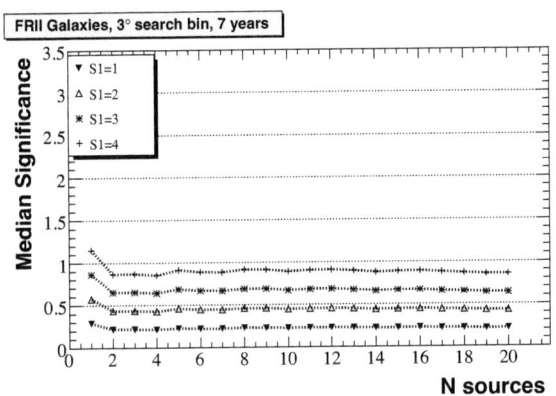

Figure B.16: Median significance of the sample of FR-II sources in dependence of the number of sources. This sample is dominated by 3C123.

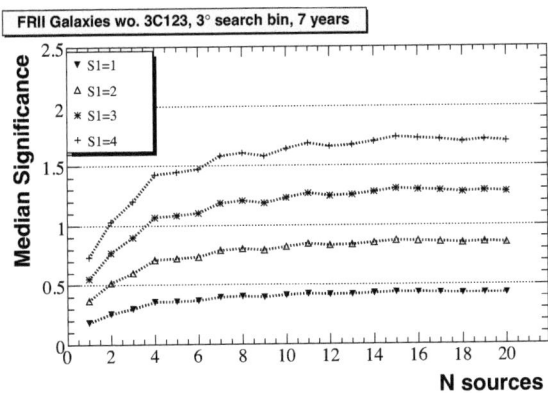

Figure B.17: The same plot without 3C123 in dependence of the number of sources, 15 sources should be stacked.

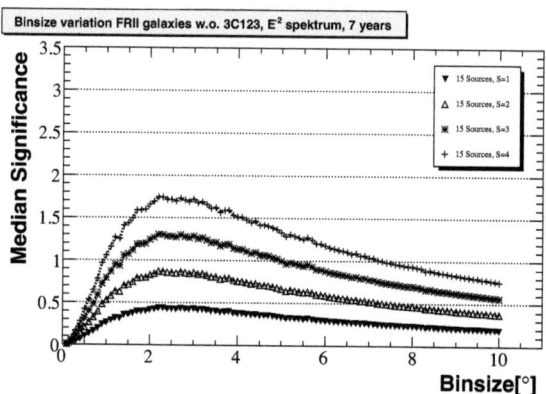

Figure B.18: The search bin variation for FR-II sources yields an optimal search of 2.4°.

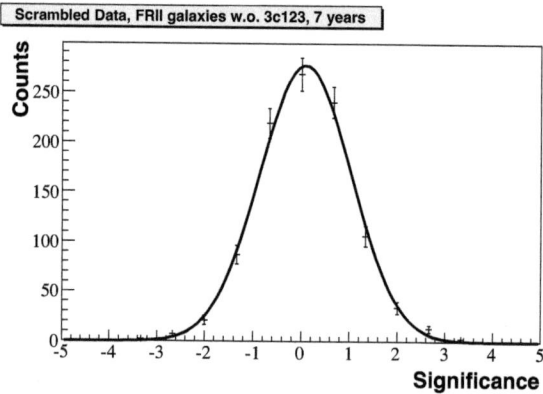

Figure B.19: The mean of the significance of the analysis of scrambled data sets is 0.

Blazars

The blazar sample was stacked with the flux measured by Fermi LAT in an energy range from 1 – 100 GeV. This sample was dominated by 0FGL J0238.6+1636. For this source Fermi LAT does not name a correlation to a previous detected source and has marked this source as variable. The optimization results in a stacking of 12 sources with a search bin of 2.4°. Relative strength can be seen in figure B.20.

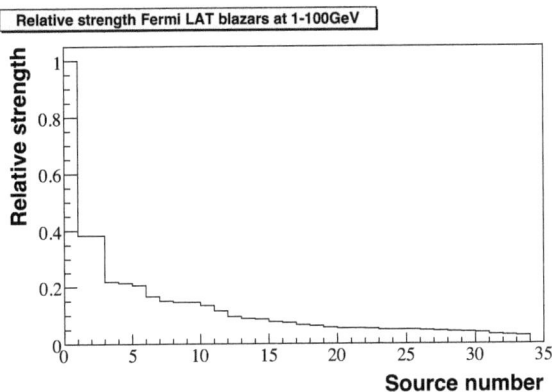

Figure B.20: The relative strength of the Fermi LAT blazars between 1 GeV and 100 GeV.

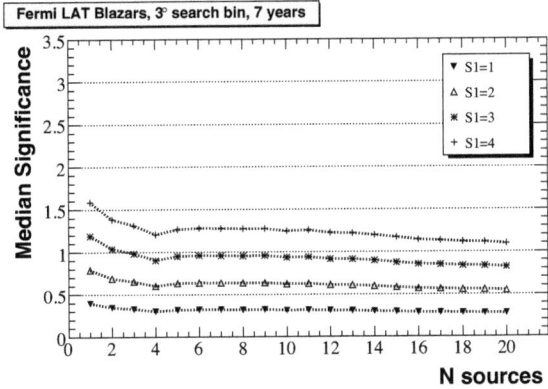

Figure B.21: The sample is dominated by the strongest source, the significance is always decreasing.

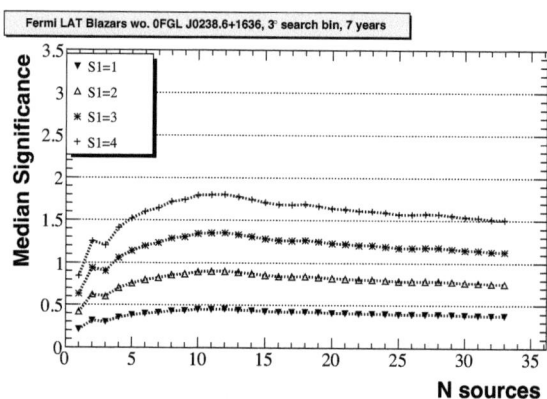

Figure B.22: The Fermi LAT blazar sample without the strongest source, 12 sources are optimal for stacking.

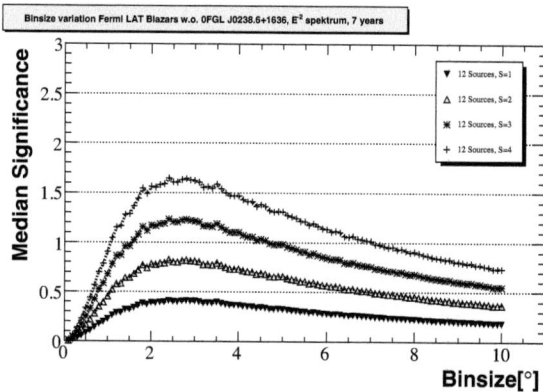

Figure B.23: The search bin variation for Fermi LAT blazars yields an optimal search of 2.4°.

B.1. Optimization plots of the AMANDA analysis

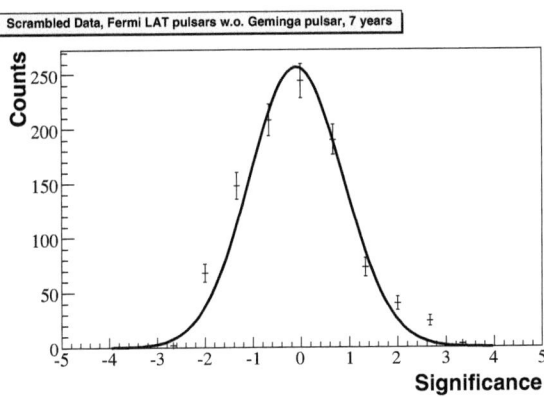

Figure B.24: The mean of the significance of the analysis of scrambled data sets is 0.

Flat spectrum radio quasars

The sample for this source class was also taken from the Fermi LAT bright sources list [Abd09]. Like the blazars it was stacked with the flux between 1 GeV and 100 GeV. This sample was dominated by the two strongest sources, 3C454.3 and PKS 1502+106. An optimization was possible after skipping these two sources and it returns a number of sources of 11 sources and a search bin of 2.6°.

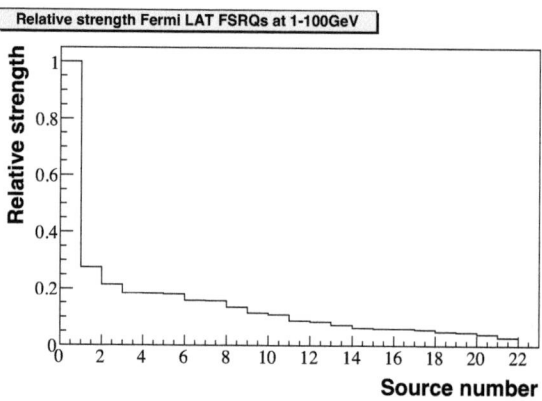

Figure B.25: The relative strength of the Fermi LAT FSRQs between 1 GeV and 100 GeV.

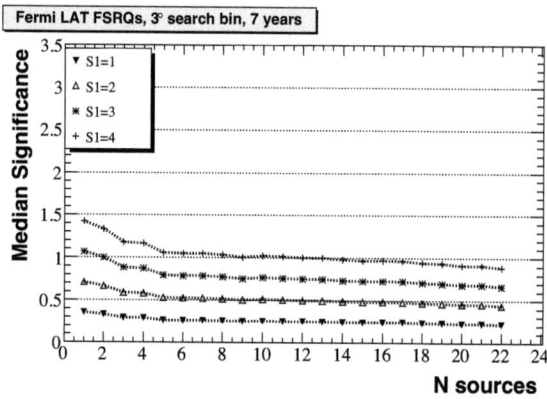

Figure B.26: The sample is dominated by the strongest source, the significance is always decreasing.

B.1. Optimization plots of the AMANDA analysis

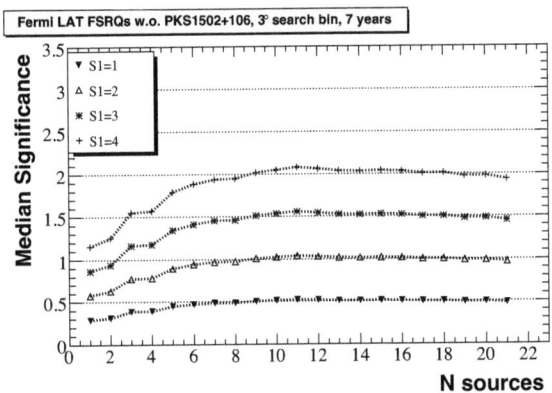

Figure B.27: The Fermi LAT blazar sample without PKS1502+106, 12 sources are optimal for stacking.

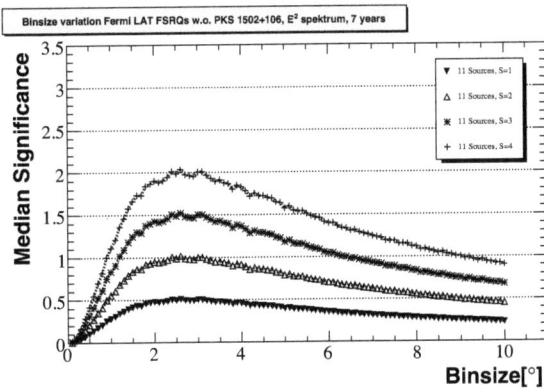

Figure B.28: The search bin variation for Fermi LAT FSRQs yields an optimal search of $2.6°$.

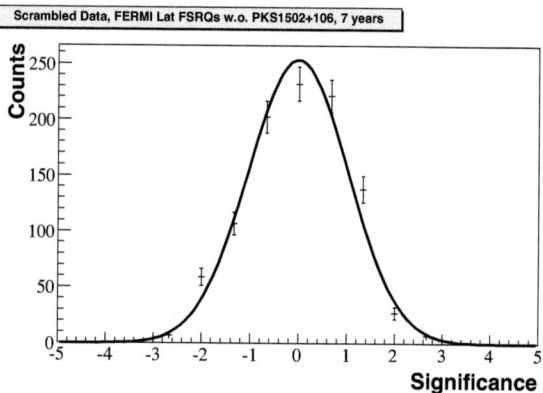

Figure B.29: For the FSRQs the analysis of scrambled data sets delivers a mean significance of 0.

Fermi LAT pulsars

Also classified and measured by Fermi LAT the pulsars were stacked with the same flux like the blazars and FSRQs measured by Fermi LAT. This sample was dominated by the Geminga pulsar which was therefore removed from the sample in order to optimize. The optimal parameters for this sample are a stacking with 4 sources and a search bin with 2.9° angular diameter. See the following figures for details.

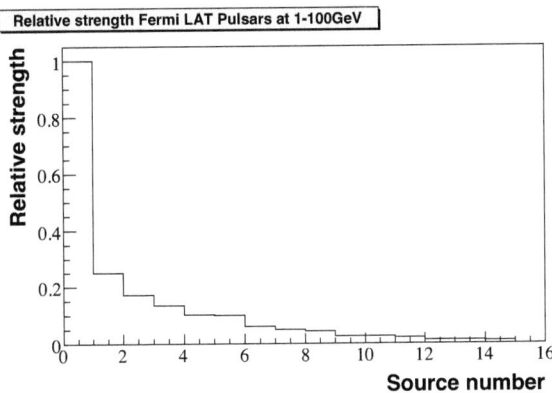

Figure B.30: The relative strength of the Fermi LAT pulsars between 1 GeV and 100 GeV.

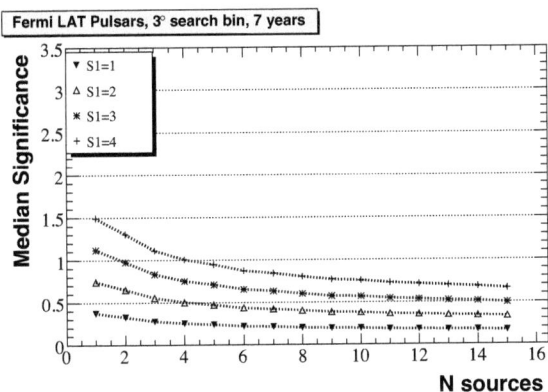

Figure B.31: The sample is dominated by the Geminga pulsar.

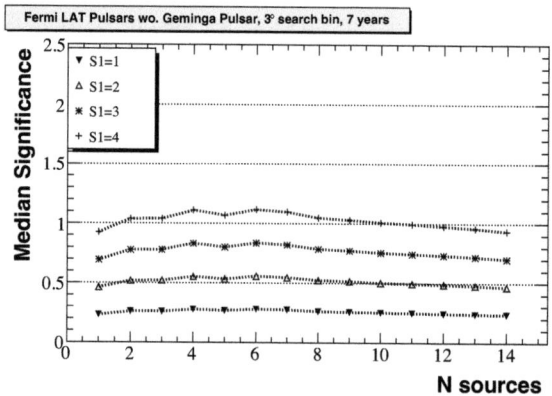

Figure B.32: The Fermi LAT pulsar sample without Geminga, 4 sources are optimal for stacking.

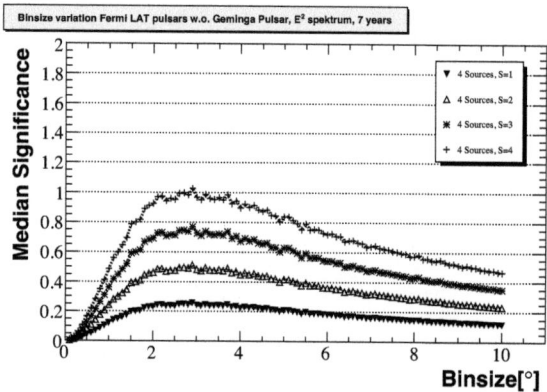

Figure B.33: The search bin variation for Fermi LAT pulsars yields an optimal search of 2.9°.

B.1. Optimization plots of the AMANDA analysis

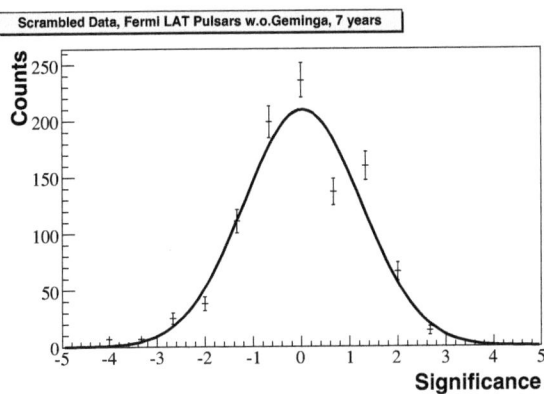

Figure B.34: The analyzed pulsars with scrambled data sets shows a mean significance of 0.

Randomized sources

The analysis method has also been tested with 1000 source catalogs with sources with randomized RA and DEC. A search bin of 2.4° has been used and 15 sources were stacked. This are the parameters obtained for the Starburst-Galaxies. The distribution of median significance of the stacked signal of the catalogs can be seen in figure B.35. As expected for randomized sources this distribution is Gaussian with a mean of 0 like in the analysis with the scrambled data sets. Similar analysis for other parameters give the same result.

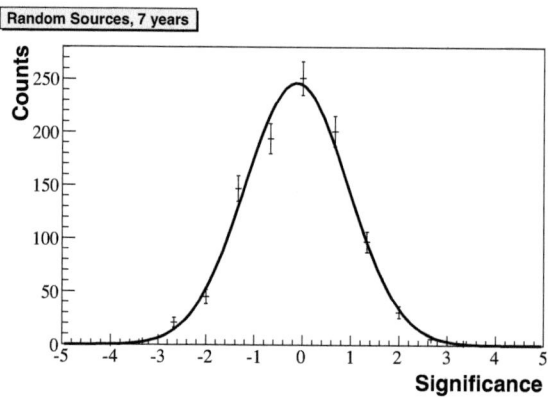

Figure B.35: Significance distribution in an analysis with 1000 randomized source catalogs.

B.2 Optimization plots of the IC-22 analysis

Starburst-Galaxies

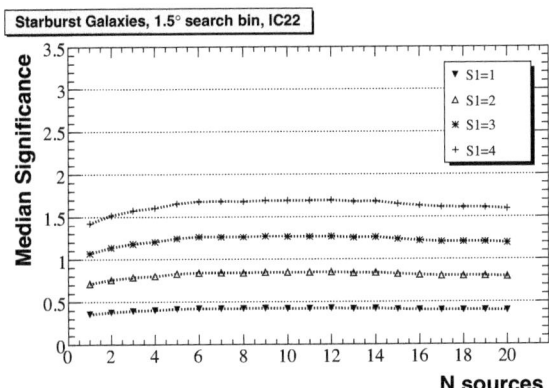

Figure B.36: Median significance of the sample of Starburst-Galaxies in dependence of the number of sources for 1, 2, 3 or 4 signal neutrinos from the strongest source. The optimal number of sources was determined to be 8 sources.

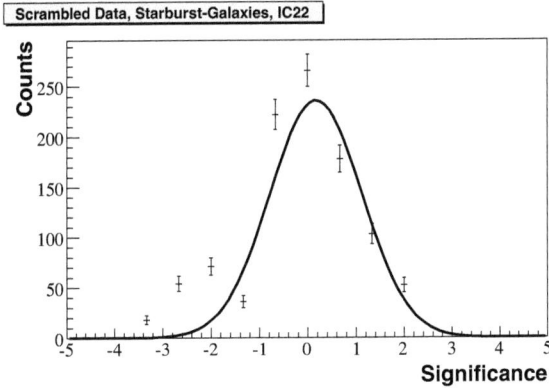

Figure B.37: Distribution of the median significance for 1000 scrambled data sets. It follows a Gaussian distribution with a mean of 0.

CSS/GPS sources

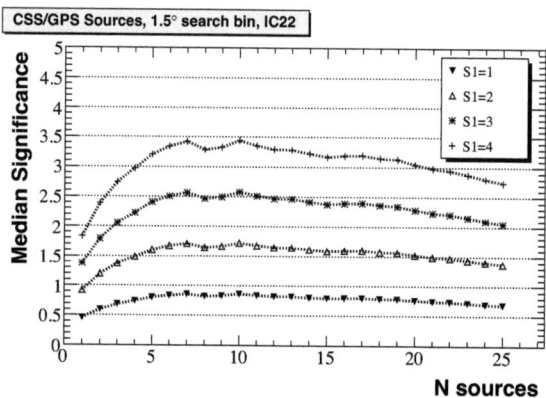

Figure B.38: Median significance for the CSS/GPS sources inn dependence of the number of sources for 1, 2, 3 or 4 signal neutrinos from the strongest source. The optimal number of sources was determined to be 7 sources.

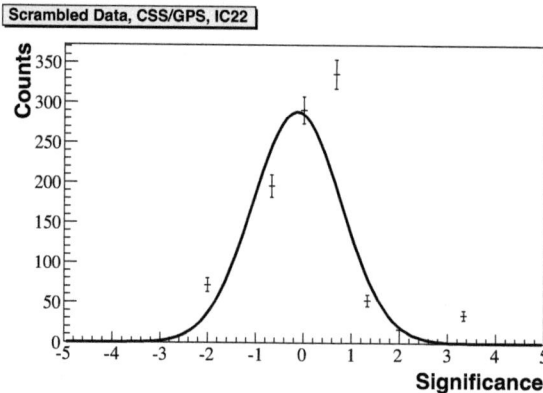

Figure B.39: Distribution of the median significance for 1000 scrambled data sets. It follows a Gaussian distribution with a mean of 0.

FR-I galaxies

B.2. Optimization plots of the IC-22 analysis

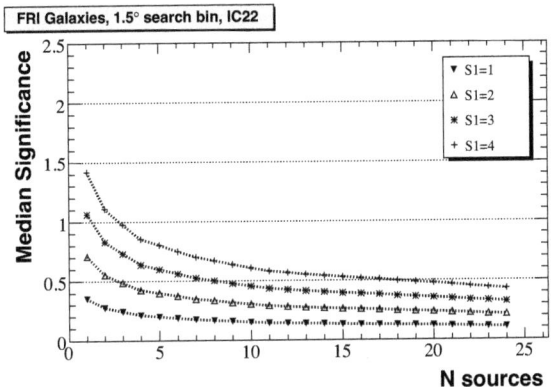

Figure B.40: Median significance of the sample of FR-I galaxies in dependence of the number of sources for 1, 2, 3 or 4 signal neutrinos from the strongest source. The sample is dominated by M87.

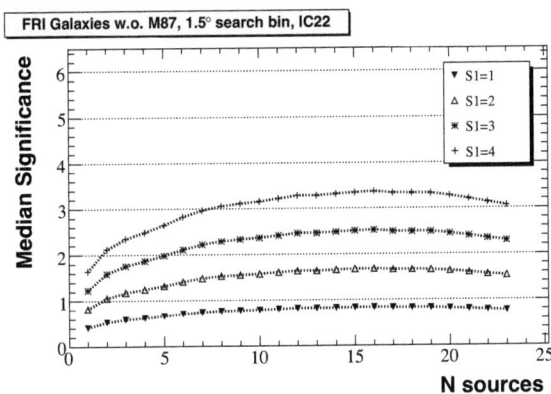

Figure B.41: Median significance of the sample of FR-I galaxies in dependence of the number of sources for 1, 2, 3 or 4 signal neutrinos from the strongest source, leaving out the dominating source. The optimal number of sources is 16.

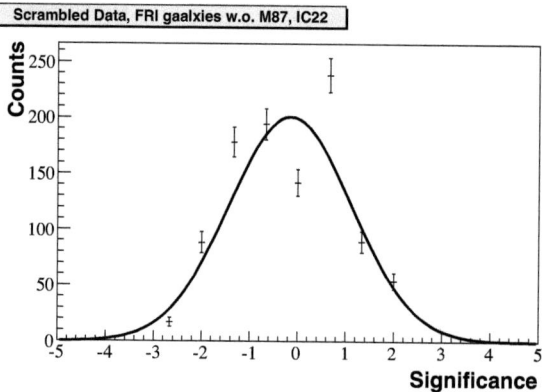

Figure B.42: Distribution of the median significance for 1000 scrambled data sets. It follows a Gaussian distribution with a mean of 0.

FR-II galaxies

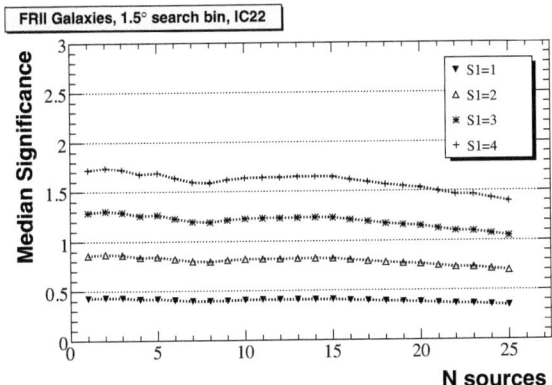

Figure B.43: Median significance of the sample of FR-II galaxies in dependence of the number of sources for $1, 2, 3$ or 4 signal neutrinos from the strongest source. The optimal number of sources was found to be 2 sources.

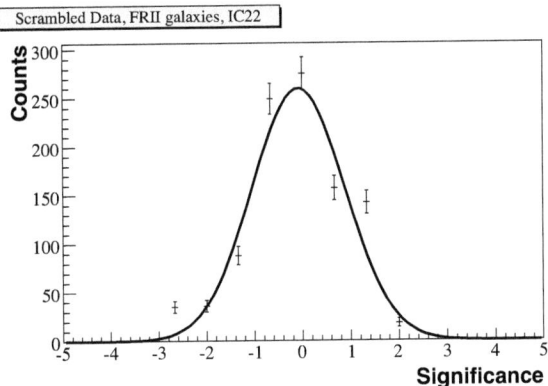

Figure B.44: Distribution of the median significance for 1000 scrambled data sets. It follows a Gaussian distribution with a mean of 0.

Fermi LAT blazars

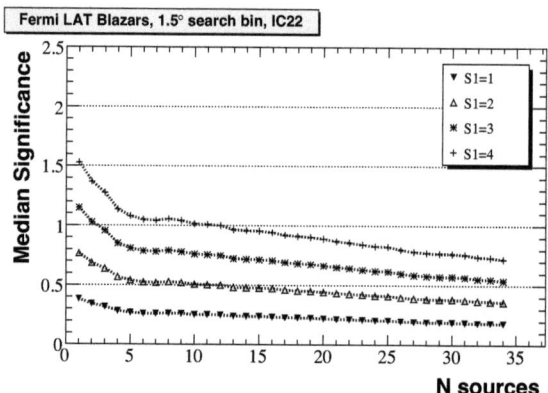

Figure B.45: Median significance of the sample of blazars in dependence of the number of sources for 1, 2, 3 or 4 signal neutrinos from the strongest source. The sample is dominated by 0FGL J0238.6+1636.

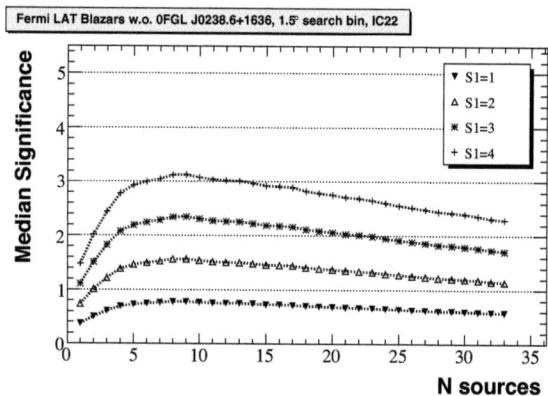

Figure B.46: Median significance of the sample of blazars in dependence of the number of sources for 1, 2, 3 or 4 signal neutrinos from the strongest source, leaving out the dominating source. The optimal number of sources is 9.

B.2. Optimization plots of the IC-22 analysis

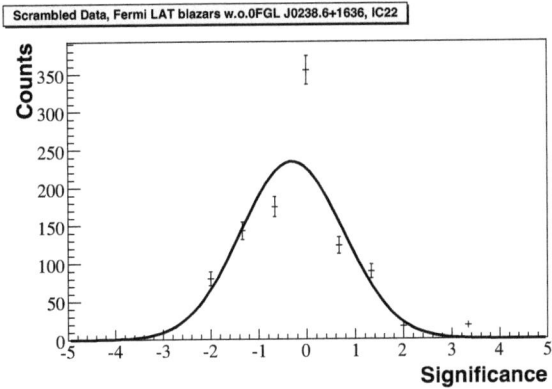

Figure B.47: Distribution of the median significance for 1000 scrambled data sets. It follows a Gaussian distribution with a mean of 0.

Fermi LAT FSRQs

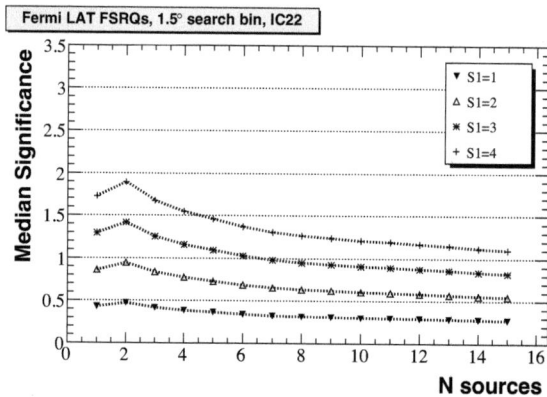

Figure B.48: Median significance of the sample of FSRQs in dependence of the number of sources for 1, 2, 3 or 4 signal neutrinos from the strongest source. The optimal number of sources is 2 sources.

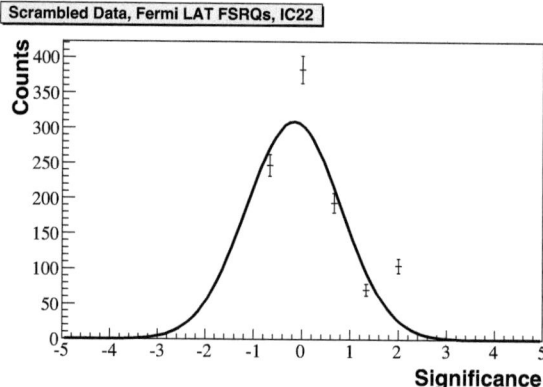

Figure B.49: Distribution of the median significance for 1000 scrambled data sets. It follows a Gaussian distribution with a mean of 0.

Fermi LAT pulsars

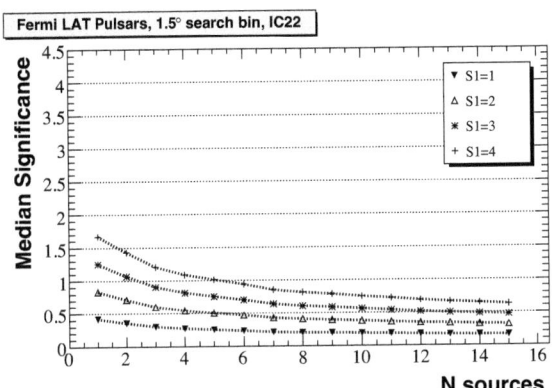

Figure B.50: Median significance of the sample of pulsars in dependence of the number of sources for $1, 2, 3$ or 4 signal neutrinos from the strongest source. The sample is dominated by the Geminga pulsar (0FGL J0634.0+1745).

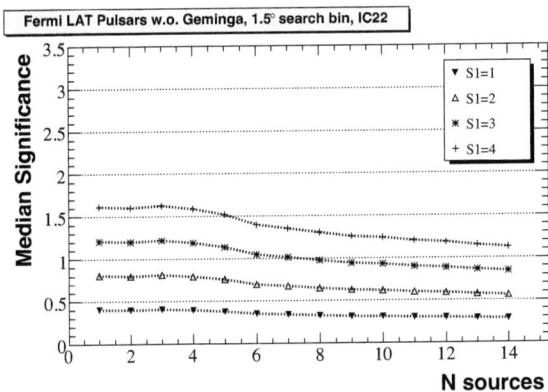

Figure B.51: Median significance of the sample of pulsars in dependence of the number of sources for $1, 2, 3$ or 4 signal neutrinos from the strongest source, leaving out the dominating source. The optimal number of sources is 3.

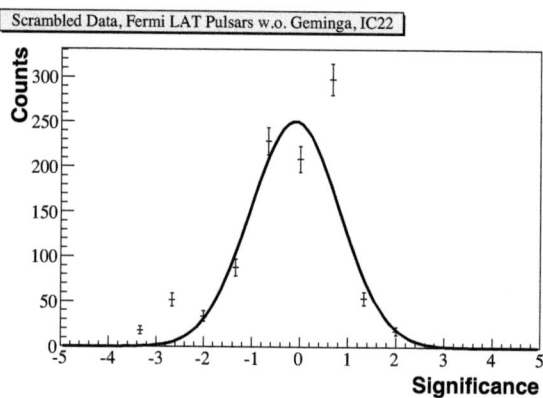

Figure B.52: Distribution of the median significance for 1000 scrambled data sets. It follows a Gaussian distribution with a mean of 0.

Appendix C

Source lists

In this appendix there are the lists of sources stacked in both analyzes, RA and DEC are in degress and use the epoch J2000.0. The positions were aquired from the NASA extragalactic database (NED), the source catalogs for the FR-I and FR-II galaxies as well as the CSS/GPS sources are from [Gro06] and references therin. The complete source catalog of Starburst-Galaxies was discussed in 3.1 and can be found in A. The source catalogs for the blazars, FSRQs and pulsars were taken from [Abd09].

C.1 Sources lists for the AMANDA analysis

Starburst-Galaxies

Source	RA [°]	DEC [°]	Flux@60 µm [Jy]
IC 342	56.7021	68.0961	180.8
NGC 2146	94.6571	78.357	146.69
Maffei2	40.4795	59.6041	135
NGC 6946	308.718	60.1539	129.78
NGC 3690	172.134	58.5622	113.05
ARP 220	233.738	23.5032	104.09
NGC 5194 (M51)	202.47	47.1952	97.42
NGC 4631	190.533	32.5415	85.4
NGC 891	35.6392	42.3491	66.46
NGC 3627 (M66)	170.063	12.9915	66.31
NGC 6606	25.7598	13.6457	65.52
NGC 2903	143.042	21.5008	60.54
NGC 3628	170.071	13.5895	54.8

Table C.1: *The Starburst-Galaxies selected for stacking with the AMANDA data.*

CSS/GPS sources

Source	RA [°]	DEC [°]	Flux@1.4 GHz [mJy]
3C147	85.6505746	49.8520094	22880.0
3C48	24.4220808	33.1597594	16018.2
3C286	202.7845329	30.5091550	14903.2
3C138	80.2911917	16.6394586	8603.3
3C309.1	224.7815996	71.6721853	7468.9
3C287	202.6570442	25.1530214	7052.6
4C12.50	206.8890067	12.2900667	5155.0

Table C.2: *The CSS/GPS selected for stacking with the AMANDA data.*

FR-I galaxies

Source	RA [°]	DEC [°]	Flux@178 MHz [mJy]
PerseusA	49.9506671	41.5116961	67.7
3C433.0	320.935559	25.069967	54.4
NGC 6166	247.160333	39.551556	51.1
3C310.0	226.238021	26.016261	45.0
3C66.0	35.6064	43.0132	35.7
NGC 1265	49.566083	41.857722	33.3
3C386.0	279.609395	17.197023	26.0
NGC 3862	176.2708708	19.6063169	26.0
UGC 11958	333.69542	13.84000	22.0
3C288.0	204.707809	38.852568	20.6
M 84	186.2655971	12.8869831	19.0
3C315.0	228.416983	26.125335	18.2
NGC 383	16.8539946	32.4125594	17.97
3C346.0	250.9524958	17.2637389	11.8

Table C.3: *The FR-I galaxies selected for stacking with the AMANDA data.*

FR-II galaxies

Source	RA [°]	DEC [°]	Flux@178 MHz [mJy]
3C295.0	212.836029	52.202512	83.5
3C196.0	123.4001379	48.2173778	68.2
3C452.0	341.453208	39.687694	54.4
3C33.0	17.220251	13.337166	54.4
3C390.3	280.5374579	79.7714242	47.5
3C98.0	59.726792	10.434167	47.2
3C20.0	10.786841	52.059387	42.9
3C219.0	140.285958	45.649278	41.2
3C234.0	150.456485	28.785919	31.4
3C61.1	35.65000	86.31889	31.2
3C79.0	47.500375	17.099528	30.5
3C330.0	242.402530	65.945446	27.8
3C427.1	316.026561	76.553199	26.6
3C47.0	24.101667	20.957500	26.4
3C388.0	281.010000	45.558250	45.5

Table C.4: *The FR-II galaxies selected for stacking with the AMANDA data.*

Fermi LAT blazars

Source	RA [°]	DEC [°]	Flux@1 GeV − 100 GeV [10^{-8}cm^{-2}s^{-1}]
0FGL J1104.5+3811	166.137	38.187	2.61
0FGL J0222.6+4302	35.653	43.043	2.61
0FGL J0722.0+7120	110.508	71.348	1.49
0FGL J1555.8+1110	238.951	11.181	1.46
0FGL J1218.0+3006	184.517	30.108	1.41
0FGL J1719.3+1746	259.830	17.769	1.14
0FGL J1221.7+2814	185.439	28.243	1.03
0FGL J1751.5+0935	267.893	9.591	1.00
0FGL J1015.2+4927	153.809	49.463	1.00
0FGL J1427.1+2347	216.794	23.785	0.92
0FGL J0818.3+4222	124.579	42.367	0.79
0FGL J0112.1+2247	18.034	22.790	0.65

Table C.5: *The Fermi LAT balzars selected for stacking with the AMANDA data.*

Fermi LAT FSRQs

Source	RA [°]	DEC [°]	Flux@1 GeV – 100 GeV [10^{-8}cm^{-2}s^{-1}]
0FGL J1229.1+0202	187.29	2.05	1.61
0FGL J0654.3+4513	103.59	45.22	1.26
0FGL J1553.4+1255	238.37	12.92	1.08
0FGL J1849.4+6706	282.37	67.1	1.07
0FGL J1522.2+3143	230.55	31.73	1.06
0FGL J1310.6+3220	197.66	32.34	0.93
0FGL J1635.2+3809	248.82	38.16	0.92
0FGL J0957.6+5522	149.42	55.38	0.79
0FGL J1015.9+0515	153.99	5.25	0.67
0FGL J1847.8+3223	281.96	32.39	0.64
0FGL J0714.2+1934	108.55	19.57	0.51

Table C.6: *The Fermi LAT FSRQs selected for stacking with the AMANDA data.*

Fermi LAT pulsars

Source	RA [°]	DEC [°]	Flux@1 GeV – 100 GeV [10^{-8}cm^{-2}s^{-1}]
0FGL J0534.6+2201(Crab)	83.65	22.02	15.40
0FGL J2021.5+4026	305.4	40.44	10.60
0FGL J1836.2+5924	279.06	59.41	8.36
0FGL J2020.8+3649	305.22	36.83	6.28

Table C.7: *The Fermi LAT pulsars selected for stacking with the AMANDA data.*

C.2 Sources lists for the IC-22 analysis

Here the sources were ranked according to a stacking parameter described in 5.4.

Starburst-Galaxies

Source	RA [°]	DEC [°]	Stack. Param.
NGC 3034 (M82)	148.968	69.6797	1
ARP 220	233.738	23.5032	0.188641
NGC 660	25.7598	13.6457	0.13568
NGC 3627 (M66)	170.063	12.9915	0.135091
IC 342	56.7021	68.0961	0.12754
NGC 4631	190.533	32.5415	0.118678
NGC 2903	143.042	21.5008	0.111758
NGC 5194 (M51)	202.47	47.1952	0.110221

Table C.8: *The Starburst-Galaxies selected for stacking with the IC-22 data.*

CSS/GPS sources

Source	RA [°]	DEC [°]	Stack. Param.
3C286	202.785	30.5092	1
3C147	85.6506	49.852	0.999272
3C138	80.2912	16.6395	0.75416
3C48	24.4221	33.1598	0.686753
3C287	202.657	25.153	0.415196
4C12.50	206.889	12.2901	0.39572
PKS 0428+20	67.7657	20.6262	0.289942

Table C.9: *The CSS/GPS sources selected for stacking with the IC-22 data.*

FR-I galaxies

Source	RA [°]	DEC [°]	Stack. Param.
3C 433.0	320.936	25.07	1
Perseus A	49.9507	41.5117	0.821729
3C 310.0	226.238	26.0163	0.810219
NGC 6166	247.16	39.5516	0.696666
3C 386.0	279.609	17.197	0.571184
NGC 3862	176.271	19.6063	0.556089
3C 66.0	35.6064	43.0132	0.524969
UGC 11958	333.695	13.84	0.513734
M 84	186.266	12.887	0.487639
NGC 1265	49.5661	41.8577	0.438249
3C 315.0	228.417	26.1253	0.40578
3C 28.0	13.9597	26.4096	0.33799
NGC 5532	214.221	10.8074	0.33715
3C 346.0	250.952	17.2637	0.333081
3C 288.0	204.708	38.8526	0.292762
NGC 7720	354.622	27.0315	0.290775

Table C.10: *The FR-I galaxies selected for stacking with the IC-22 data.*

FR-II galaxies

Source	RA [°]	DEC [°]	Stack. Param.
3C 123.0	69.2682	29.6705	1
3C 33.0	17.2203	13.3372	0.3391

Table C.11: *The FR-II galaxies selected for stacking with the IC-22 data.*

Fermi LAT blazars

Source	RA [°]	DEC [°]	Stack. Param.
0FGL J1104.5+3811	166.137	38.187	1
0FGL J0222.6+4302	35.653	43.043	0.692387
0FGL J1555.8+1110	238.951	11.181	0.666191
0FGL J1751.5+0935	267.893	9.591	0.63702
0FGL J1719.3+1746	259.83	17.769	0.616565
0FGL J1218.0+3006	184.517	30.108	0.494447
0FGL J1427.1+2347	216.794	23.785	0.389905
0FGL J1221.7+2814	185.439	28.243	0.387164
0FGL J1015.2+4927	153.809	49.463	0.312798

Table C.12: *The Fermi LAT blazars selected for stacking with the IC-22 data.*

Fermi LAT FSRQs

Source	RA [°]	DEC [°]	Stack. Param.
0FGL J2254.0+1609	343.5	16.15	1
0FGL J1504.4+1030	226.12	10.51	0.603106

Table C.13: *The Fermi LAT FSRQs selected for stacking with the IC-22 data.*

Fermi LAT pulsars

Source	RA [°]	DEC [°]	Stack. Param.
0FGL J0534.6+2201(Crab)	83.65	22.02	1
0FGL J2021.5+4026	305.4	40.44	0.489727
0FGL J1907.5+0602	286.89	6.03	0.348537

Table C.14: *The Fermi LAT pulsars selected for stacking with the IC-22 data.*

List of Figures

2.1	Cosmic ray spectrum	4
2.2	Fermi acceleration	6
2.3	The Hillas diagram	7
2.4	An AGN	9
2.5	The AGN classification scheme	11
2.6	Cosmic ray propagation and detection	13
2.7	An air shower	15
2.8	The Čerenkov effect	17
2.9	Neutrino spectrum	25
3.1	Luminosity - distance diagram (FIR)	34
3.2	Luminosity - distance diagram (Radio)	35
3.3	Log(N)-Log(S) presentation	36
3.4	Radio power versus FIR luminosity	37
3.5	FIR to radio flux ratio	38
3.6	Radio versus X-ray index	39
3.7	Expected diffuse neutrino flux from SNRs in Starburst-Galaxies	41
3.8	Number of GRBs in Starburst-Galaxies	43
3.9	Number of events from Starburst-Galaxies in IceCube	45
4.1	The IceCube detector	49
5.1	Zenith band	53
5.2	RA distributions	53
5.3	# of sources Starburst-Galaxies without M 82	57
5.4	Bin size variation Fermi Starburst-Galaxies	57
5.5	Significance distributions scrambled data/random sources	58
5.6	Effective area IC-22	60
5.7	AMANDA and IC-22 zenith angle distributions	61
5.8	Analysis results AMANDA	64
5.9	Analysis results IC-22	65
B.1	Rel. strength Starburst-Galaxies	89

B.2	# of sources Starburst-Galaxies	90
B.3	# of sources Starburst-Galaxies without M 82	90
B.4	Bin size variation Starburst-Galaxies	91
B.5	Scrambled data Starburst-Galaxies	91
B.6	Rel. strength CSS/GPS	92
B.7	# of sources CSS/GPS	92
B.8	Bin size variation CSS/GPS	93
B.9	Scrambled data CSS/GPS	93
B.10	Rel. strength FR-I galaxies	94
B.11	Rel. strength FR-II galaxies	94
B.12	# of sources FR-I	95
B.13	# of sources FR-I without M 87	95
B.14	Search bin variation FR-I	96
B.15	Scrambled data FR-I	96
B.16	# of sources FR-II	97
B.17	# of sources FR-I without 3C123.0	97
B.18	Bin size variation FR-II	98
B.19	Scrambled data FR-II	98
B.20	Rel. strength Fermi LAT blazars	99
B.21	# of sources Fermi LAT blazars	99
B.22	# of sources Fermi LAT blazars without strongest source	100
B.23	Bin size variation Fermi LAT blazars	100
B.24	Scrambled data Fermi LAT blazars	101
B.25	Rel. strength Fermi LAT FSRQs	102
B.26	# of sources Fermi LAT FSRQs	102
B.27	# of sources Fermi LAT FSRQs without PKS 1502+106	103
B.28	Bin size variation Fermi LAT FSRQs	103
B.29	Scrambled data Fermi LAT FSRQs	104
B.30	Rel. strength Fermi LAT pulsars	105
B.31	# of sources Fermi LAT pulsars	105
B.32	# of sources Fermi LAT pulsars without Geminga	106
B.33	Search bin variation Fermi LAT blazars	106
B.34	Scrambled data Fermi LAT pulsars	107
B.35	Randomized source catalogs	108
B.36	# of sources Starburst-Galaxies IC-22	109
B.37	Scrambled data Starburst-Galaxies IC-22	109
B.38	# of sources CSS/GPS IC-22	110
B.39	Scrambled data CSS/GPS IC-22	110
B.40	# of sources FR-I IC-22	111

- B.41 # of sources FR-I IC-22 without M 87 111
- B.42 Scrambled data FR-I IC-22 112
- B.43 # of sources FR-II IC-22 113
- B.44 Scrambled data FR-II IC-22 113
- B.45 # of sources Fermi LAT blazars IC-22 114
- B.46 # of sources Fermi LAT blazars IC-22 without strongest source 114
- B.47 Scrambled data Fermi LAT blazars IC-22 115
- B.48 # of sources Fermi LAT FSRQs IC-22 116
- B.49 Scrambled data Fermi LAT FSRQs IC-22 116
- B.50 # of sources Fermi LAT pulsars IC-22 117
- B.51 # of sources Fermi LAT pulsars without Geminga IC-22 117
- B.52 Scrambled data Fermi LAT pulsars IC-22 118

List of Tables

2.1	Neutrino models	26
4.1	Stages of IceCube	49
5.1	Optimization results AMANDA	59
5.2	Optimization results IC-22	62
5.3	Results AMANDA	63
5.4	Results IC-22	63
A.1	Catalog: General data	69
A.2	Catalog: Radio data	74
A.3	Catalog: FIR data	80
A.4	Catalog: X-ray data	85
C.1	Starburst-Galaxies stacked AMANDA	120
C.2	CSS/GPS stacked AMANDA	121
C.3	FR-I galaxies stacked AMANDA	121
C.4	FR-II galaxies stacked AMANDA	122
C.5	Fermi LAT blazar stacked AMANDA	123
C.6	Fermi LAT FSRQs stacked AMANDA	124
C.7	Fermi LAT pulsars stacked AMANDA	124
C.8	Starburst-Galaxies stacked IC-22	125
C.9	CSS/GPS stacked IC-22	125
C.10	FR-I stacked IC-22	126
C.11	FR-II stacked IC-22	126
C.12	Fermi LAT blazars stacked IC-22	127
C.13	Fermi LAT FSRQs stacked IC-22	127
C.14	Fermi LAT pulsars stacked IC-22	127

Bibliography

[A+93] D. E. Alexandreas et al. Point source search techniques in ultra high energy gamma ray astronomy. *Nuclear Instruments and Methods in Physics Research A*, 328:570–577, May 1993.

[A+04] J. Ahrens (IceCube Collaboration) et al. Sensitivity of the IceCube detector to astrophysical sources of high energy muon neutrinos. *Astroparticle Physics*, 20:507, February 2004.

[A+05] M. Altmann (GNO Collaboration) et al. Complete results for five years of GNO solar neutrino observations. *Physics Letters B*, 616:174–190, June 2005.

[A+06] A. Achterberg (Icecube Collaboration) et al. First year performance of the IceCube neutrino telescope. *Astroparticle Physics*, 26:155–173, October 2006.

[A+07] A. Achterberg (IceCube Collaboration) et al. Five years of searches for point sources of astrophysical neutrinos with the AMANDA-II neutrino telescope. *Physical Review D*, 75(10):102001–+, May 2007.

[A+08a] J. Abraham et al. Observation of the Suppression of the Flux of Cosmic Rays above $4 \cdot 10^{19} eV$. *Physical Review Letters*, 101(6):061101–+, August 2008.

[A+08b] E. Aliu (MAGIC Collaboration) et al. Observation of Pulsed γ-Rays Above 25 GeV from the Crab Pulsar with MAGIC. *Science*, 322:1221–, November 2008.

[A+09a] R. Abbasi (IceCube Collaboration) et al. First Neutrino Point-Source Results from the 22 String Icecube Detector. *Astrophysical Journal Letters*, 701:L47–L51, August 2009.

[A+09b] R. Abbasi (IceCube Collaboration) et al. Search for point sources of high energy neutrinos with final data from AMANDA-II. *Physical Review D*, 79(6):062001–+, March 2009.

[A+09c] R. Abbasi (IceCube Collaboration) et al. The IceCube Yellow Book, 2009. in preparation.

[A+09d] A. A. Abdo (Fermi LAT Collaboration) et al. Detection of Gamma-Ray Emission from the Starburst Galaxies M 82 and NGC 253 with the Large Area Telescope on Fermi. *ArXiv e-prints: 0911.5327*, November 2009.

[A+09e]   J. Abraham (Pierre Auger Collaboration) et al. Operations of and Future Plans for the Pierre Auger Observatory. *ArXiv e-prints: 0906.2354*, June 2009.

[A+09f]   V. A. Acciari (VERITAS Collaboration) et al. A connection between star formation activity and cosmic rays in the starburst galaxy M 82. *ArXiv e-prints: 0911.0873*, November 2009.

[A+09g]   F. Acero (H.E.S.S. Collaboration) et al. Detection of gamma rays from a starburst galaxy. *Science*, 326:1080–1082, 2009.

[A+09h]   H. Anderhub et al. MAGIC Collaboration: Contributions to the 31st International Cosmic Ray Conference (ICRC 2009). *ArXiv e-prints: 0907.0843*, July 2009.

[AAB+06] G. Aggouras, E. G. Anassontzis, Ball, et al. Recent results from NESTOR. *Nuclear Instruments and Methods in Physics Research A*, 567:452–456, November 2006.

[Abd09]   A. A. Abdo. Bright AGN Source List from the First Three Months of the Fermi Large Area Telescope All-Sky Survey. *ArXiv e-prints: 0902.1559*, February 2009.

[AD01]   A. Atoyan and C. D. Dermer. High-Energy Neutrinos from Photomeson Processes in Blazars. *Physical Review Letters*, 87(22):221102–+, November 2001.

[AGHW04] L. A. Anchordoqui, H. Goldberg, F. Halzen, and T. J. Weiler. Neutrino bursts from Fanaroff Riley I radio galaxies. *Physics Letters B*, 600:202–207, October 2004.

[AM04]   J. Alvarez-Muñiz and P. Mészáros. High energy neutrinos from radio-quiet active galactic nuclei. *Physical Review D*, 70(12):123001–+, December 2004.

[And33]   Carl D. Anderson. The positive electron. *Physical Review*, 43(6):491–494, March 1933.

[AP+07]   J. Abraham (Pierre Auger Collaboration) et al. Correlation of the Highest-Energy Cosmic Rays with Nearby Extragalactic Objects. *Science*, 318:938–, November 2007.

[AP+08]   J. Abraham (Pierre Auger Collaboration) et al. Erratum to 'Correlation of the highest-energy cosmic rays with the positions of nearby active galactic nuclei' [Astroparticle Physics 29(3) (2008) 188 204]. *Astroparticle Physics*, 30:45–45, August 2008.

[AR90]   A. Wright and R. Otrupcek. Parkes catalogue. *Australia Telescope National Facility*, 1990.

[AR06]   L. Armus and W. T. Reach. The Spitzer Space Telescope: New Views of the Cosmos. In L. Armus and W.T. Reach, editor, *Astronomical Society of the Pacific Conference Series*, volume 357 of *Astronomical Society of the Pacific Conference Series*, December 2006.

[Ask62]   Gurgen A. Askaryan. Excess Negative Charge of the Electron-Photon Shower and Coherent Radiation Originating from It. Radio Recording of Showers under the

Ground and on the Moon. *Journal of the Physical Society of Japan Supplement*, 17:C257+, 1962.

[B+73] A. Boksenberg et al. The ultra-violet sky-survey telescope in the TD-IA satellite. *Monthly Notices of the Royal Astronomical Society*, 163:291–322, 1973.

[B+03] E. Berger et al. A common origin for cosmic explosions inferred from calorimetry of GRB030329. *Nature*, 426:154, November 2003.

[B+08] D. R. Bergman (HiRes Collaboration) et al. Observation of the GZK Cutoff by the HiRes Experiment. In *International Cosmic Ray Conference*, volume 4 of *International Cosmic Ray Conference*, pages 451–454, 2008.

[B+09a] P. L. Biermann et al. Active Galactic Nuclei: Sources for ultra high energy cosmic rays? *Nuclear Physics B Proceedings Supplements*, 190:61, May 2009.

[B+09b] A. Brunthaler et al. Discovery of a bright radio transient in M 82: a new radio supernova? *Astronomy & Astrophysics*, 499:L17–L20, May 2009.

[BB09] J. K. Becker and P. L. Biermann. Neutrinos from active black holes, sources of ultra high energy cosmic rays. *Astroparticle Physics*, 31:138–148, March 2009.

[BBB+04] H. Bravo-Alfaro, E. Brinks, A. J. Baker, F. Walter, and D. Kunth. H I and CO in Blue Compact Dwarf Galaxies: Haro 2 and Haro 4. *Astronomical Journal*, 127:264–278, January 2004.

[BBDK09] J. K. Becker, P. L. Biermann, J. Dreyer, and T. M. Kneiske. Cosmic Rays VI - Starburst galaxies at multiwavelengths. *ArXiv e-prints: 0901.1775*, January 2009. Submitted to A&A.

[BBM05] W. Bednarek, G. F. Burgio, and T. Montaruli. Galactic discrete sources of high energy neutrinos. *New Astronomy Review*, 49:1–21, February 2005.

[BBR05a] J. K. Becker, P. L. Biermann, and W. Rhode. A source property based estimate of the neutrino flux from blazars and steep spectrum sources. In *International Cosmic Ray Conference*, volume 5 of *Proceedings of the 29th International Cosmic Ray Conference. August 3-10, 2005, Pune, India*, pages 9–+, 2005.

[BBR05b] J. K. Becker, P. L. Biermann, and W. Rhode. The diffuse neutrino flux from FR-II radio galaxies and blazars: A source property based estimate. *Astroparticle Physics*, 23:355–368, May 2005.

[Bec04] Julia K. Becker. Calculation of the agn neutrino flux and of event rates for large volume neutrino telescopes. Master's thesis, Universität Wuppertal, 2004. WU D 04-06.

[Bec07] Julia K. Becker. *Neutrinos On The Rocks—On the phenomenology of potential astrophysical neutrino sources*. PhD thesis, Universität Dortmund, 2007.

[Bec08]   Julia K. Becker. High-energy neutrinos in the context of multimessenger astrophysics. *Physics Reports*, 458:173–246, March 2008.

[Bed01]   W. Bednarek. Extragalactic neutrino background from very young pulsars surrounded by supernova envelopes. *Astronomy & Astrophysics*, 378:L49–L52, October 2001.

[BEW85]   P. L. Biermann, A. Eckart, and A. Witzel. Radio source counts below 100 micro-Jy predicted from IRAS 60 micron counts. *Astronomy & Astrophysics*, 142:L23, January 1985.

[BNH+88]   C. A. Beichman, G. Neugebauer, Habing, et al., editors. *Infrared astronomical satellite (IRAS) catalogs and atlases. Volume 1: Explanatory supplement*, volume 1, 1988.

[Bro09]   A. M. Brown. The ANTARES neutrino telescope: Status and first results. In C. Balazs & F. Wang, editor, *American Institute of Physics Conference Series*, volume 1178 of *American Institute of Physics Conference Series*, pages 76–82, October 2009.

[BS87]   P. L. Biermann and P. A. Strittmatter. Synchrotron emission from shock waves in active galactic nuclei. *The Astrophysical Journal*, 322:643–649, November 1987.

[BSB94]   W. Brinkmann, J. Siebert, and T. Boller. The X-ray AGN content of the Molonglo 408 MHz Survey: Bulk properties of previously optically identified sources. *Astronomy & Astrophysics*, 281:355–374, January 1994.

[BSHR06]   J. K. Becker, M. Stamatikos, F. Halzen, and W. Rhode. Coincident GRB neutrino flux predictions: Implications for experimental UHE neutrino physics. *Astroparticle Physics*, 25:118–128, March 2006.

[Bur56]   Geoffrey R. Burbidge. On Synchrotron Radiation from Messier 87. *The Astrophysical Journal*, 124:416–+, September 1956.

[C+08]   M. Casolino et al. Launch of the space experiment PAMELA. *Advances in Space Research*, 42:455–466, August 2008.

[CCB02]   J. J. Condon, W. D. Cotton, and J. J. Broderick. Radio Sources and Star Formation in the Local Universe. *Astronomical Journal*, 124:675–689, August 2002.

[CCBD83]   J. J. Condon, M. A. Condon, J. J. Broderick, and M. M. Davis. Optical identifications of flat-spectrum radio sources. *Astronomical Journal*, 88:20–36, January 1983.

[CCM99]   G. W. Collins, II, W. P. Claspy, and J. C. Martin. A Reinterpretation of Historical References to the Supernova of A.D. 1054. *The Publications of the Astronomical Society of the Pacific*, 111:871–880, July 1999.

Bibliography

[CD73] T. L. Cline and U. D. Desai. Search for Brief Celestial X-ray Bursts. In *International Cosmic Ray Conference*, volume 1 of *International Cosmic Ray Conference*, pages 80–+, 1973.

[Cir08] Vanessa Cirkel-Bartelt. History of astroparticle physics and its components. *Living Reviews in Relativity*, 11(2), 2008.

[CKB89] R. Chini, E. Kreysa, and P. L. Biermann. The nature of radio-quiet quasars. *Astronomy & Astrophysics*, 219:87, July 1989.

[CO96] Bradly C. Caroll and Dale A. Ostlie. *An Introduction to Modern Astrophysics*. Addison-Wesley Publishing Company Inc., 1996.

[Col09] IceCube Collaboration. Icecube gallery, December 2009.

[Con83] J. J. Condon. Strong radio sources in bright spiral galaxies. III - Disk emission. *The Astrophysical Journal Supplement Series*, 53:459–495, October 1983.

[CT01] E. Cappellaro and M. Turatto. Supernova Types and Rates. In D. Vanbeveren, editor, *The Influence of Binaries on Stellar Population Studies*, volume 264 of *Astrophysics and Space Science Library*, page 199, 2001. arXiv:astro-ph/0012455.

[CTR05] A. N. Cillis, D. F. Torres, and O. Reimer. EGRET Upper Limits and Stacking Searches of Gamma-Ray Observations of Luminous and Ultraluminous Infrared Galaxies. *The Astrophysical Journal*, 621:139, March 2005.

[CWY+09] Y.-M. Chen, J.-M. Wang, C.-S. Yan, C. Hu, and S. Zhang. The Starburst-Active Galactic Nucleus Connection: The Role of Young Stellar Populations in Fueling Supermassive Black Holes. *Astrophysical Journal Letters*, 695:L130–L133, April 2009.

[D+95] K. Daum (Fréjus Collaboration) et al. Determination of the atmospheric neutrino spectra with the Fréjus detector. *Zeitschrift für Physik C Particles and Fields*, 66:417–428, 1995.

[D+01] S. G. Djorgovski et al. The Afterglow and the Host Galaxy of the Dark Burst GRB 970828. *The Astrophysical Journal*, 562:654–663, December 2001.

[Dar49] Charles Darwin. Source of the Cosmic Rays. *Nature*, 164:1112–1114, December 1949.

[DC78] L. L. Dressel and J. J. Condon. The Arecibo 2380 MHz survey of bright galaxies. *The Astrophysical Journal Supplement Series*, 36:53–75, January 1978.

[Des09] F. Descamps for the IceCube Collaboration. Acoustic detection of high energy neutrinos in ice: Status and results from the South Pole Acoustic Test Setup. *ArXiv e-prints: 0908.3251*, August 2009. Proceedings of the 31st ICRC Łódź, Poland.

[Dis09] C. Distefano for the NEMO Collaboration. Status of NEMO: results from the NEMO Phase-1 detector. *ArXiv e-prints: 0901.1252*, January 2009.

[Duy42] J. J. L. Duyvendak. Further Data Bearing on the Identification of the Crab Nebula with the Supernova of 1054 A.D. Part I. The Ancient Oriental Chronicles. *The Publications of the Astronomical Society of the Pacific*, 54:91–94, April 1942.

[DW77] M. J. Disney and J. V. Wall. A 5-GHz survey of bright Southern elliptical and S0 galaxies. *Monthly Notices of the Royal Astronomical Society*, 179:235–254, April 1977.

[EG07] Julius Elster and Hans Geitel. *Über die Radioaktivität der Erdsubstanz und ihre mögliche Beziehung zur Erdwärme*. Wolfenbüttel Heckner, 1907.

[F+03] S. Fukuda (The Super-KAMIOKANDE Collaboration) et al. The Super-Kamiokande detector. *Nuclear Instruments and Methods in Physics Research A*, 501:418–462, April 2003.

[FB95] H. Falcke and P. L. Biermann. The jet-disk symbiosis. I. Radio to X-ray emission models for quasars. *Astronomy & Astrophysics*, 293:665–682, January 1995.

[FB99] H. Falcke and P. L. Biermann. The jet/disk symbiosis. III. What the radio cores in GRS 1915+105, NGC 4258, M 81 and SGR A* tell us about accreting black holes. *Astronomy & Astrophysics*, 342:49–56, February 1999.

[FC98] G. J. Feldman and R. D. Cousins. Unified approach to the classical statistical analysis of small signals. *Physical Review D*, 57:3873–3889, April 1998.

[Fer49] E. Fermi. On the Origin of the Cosmic Radiation. *Physical Review*, 75:1169–1174, April 1949.

[Fer54] Enrico Fermi. Galactic Magnetic Fields and the Origin of Cosmic Radiation. *The Astrophysical Journal*, 119:1–+, January 1954.

[FKT92] G. Fabbiano, D.-W. Kim, and G. Trinchieri. An X-ray catalog and atlas of galaxies. *The Astrophysical Journal Supplement Series*, 80:531–644, June 1992.

[FMB95] H. Falcke, M. A. Malkan, and P. L. Biermann. The jet-disk symbiosis. II.Interpreting the radio/UV correlations in quasars. *Astronomy & Astrophysics*, 298:375–+, June 1995.

[G+98] T. J. Galama et al. An unusual supernova in the error box of the γ-ray burst of 25 April 1998. *Nature*, 395:670–672, October 1998.

[G+04] D. Guetta et al. Neutrinos from individual gamma-ray bursts in the BATSE catalog. *Astroparticle Physics*, 20:429–455, January 2004.

[GA08] A. Groß and J. L. B. Alba. Search for sources of astrophysical neutrinos with IceCube. In F. A. Aharonian, W. Hofmann, & F. Rieger, editor, *American Insti-*

tute of Physics Conference Series, volume 1085 of *American Institute of Physics Conference Series*, pages 779–782, December 2008.

[Gai90] Thomas K. Gaisser. *Cosmic Rays and Particles*. Cambridge University Press, 1990.

[GAO+06] J. F. Gallimore, D. J. Axon, C. P. O'Dea, S. A. Baum, and A. Pedlar. A Survey of Kiloparsec-Scale Radio Outflows in Radio-Quiet Active Galactic Nuclei. *Astronomical Journal*, 132:546–569, August 2006.

[GBB+07] P. W. Gorham, S. W. Barwick, Beatty, et al. Observations of the Askaryan Effect in Ice. *Physical Review Letters*, 99(17):171101–+, October 2007.

[GCG+06] J. Gorosabel, A. J. Castro-Tirado, S. Guziy, et al. The short-duration GRB 050724 host galaxy in the context of the long-duration GRB hosts. *Astronomy & Astrophysics*, 450:87–92, April 2006.

[GHA+04] D. Guetta, D. Hooper, J. Alvarez-Mun~Iz, F. Halzen, and E. Reuveni. Neutrinos from individual gamma-ray bursts in the BATSE catalog. *Astroparticle Physics*, 20:429–455, January 2004.

[GMP05] M. Guainazzi, G. Matt, and G. C. Perola. X-ray obscuration and obscured AGN in the local universe. *Astronomy & Astrophysics*, 444:119–132, December 2005.

[Gre66] Kenneth Ingvar Greisen. End to the Cosmic-Ray Spectrum? *Physical Review Letters*, 16, 1966.

[Gro06] Andreas Groß. *Search for Higgh Energy Neutrinos from Generic AGN classes with AMANDA-II*. PhD thesis, Universität Dortmund, 2006.

[Gru93] Claus Grupen. *Teilchendetektoren*. Bibliographisches Institut & F.A. Brockhaus AG, 1993.

[GS04a] Y. Gao and P. M. Solomon. HCN Survey of Normal Spiral, Infrared-luminous, and Ultraluminous Galaxies. *The Astrophysical Journal Supplement Series*, 152:63–80, May 2004.

[GS04b] Y. Gao and P. M. Solomon. The Star Formation Rate and Dense Molecular Gas in Galaxies. *The Astrophysical Journal*, 606:271–290, May 2004.

[GSF+05] P. W. Gorham, D. Saltzberg, Field, et al. Accelerator measurements of the Askaryan effect in rock salt: A roadmap toward teraton underground neutrino detectors. *Physical Review D*, 72(2):023002–+, July 2005.

[Güd09] M. Güdel. A decade of X-ray astronomy with XMM-Newton. Commentary on: Jansen F., Lumb D., Altieri B., et al., 2001, A&A, 365, L1; Strüder L., Briel U., Dennerl K., et al., 2001, A&A, 365, L18; Turner M. J. L., Abbey A., Arnaud M., et al., 2001, A&A, 365, L27. *Astronomy & Astrophysics*, 500:595–596, June 2009.

[GWBE94] M. R. Griffith, A. E. Wright, B. F. Burke, and R. D. Ekers. The Parkes-MIT-NRAO (PMN) surveys. 3: Source catalog for the tropical survey ($-29° < \delta < -9.5°$). *The Astrophysical Journal Supplement Series*, 90:179–295, January 1994.

[GWBE95] M. R. Griffith, A. E. Wright, B. F. Burke, and R. D. Ekers. The Parkes-MIT-NRAO (PMN) surveys. 6: Source catalog for the equatorial survey ($-9.5° < \delta < +10.0°$). *The Astrophysical Journal Supplement Series*, 97:347–453, April 1995.

[H+88] G. Helou et al. IRAS observations of galaxies in the Virgo cluster area. *The Astrophysical Journal Supplement Series*, 68:151, October 1988.

[Hel87] D. Helfand. Bang - The supernova of 1987. *Physics Today*, 40:24–32, August 1987.

[Hes11] Victor Franz Hess. Über die Absorption der g-Strahlung in der Atmosphäre. *Physikalische Zeitschrift*, 12(22,23):998–1001, 1911.

[Hes12] Viktor Franz Hess. Über Beobachtungen der durchdringenden Strahlung bei sieben Freiballonfahrten. *Physikalische Zeitschrift*, 13:1084–1091, 1912.

[HF03] T. Huege and H. Falcke. Radio emission from cosmic ray air showers. Coherent geosynchrotron radiation. *Astronomy & Astrophysics*, 412:19–34, December 2003.

[Hil84] Anthony Michael Hillas. The Origin of Ultra-High-Energy Cosmic Rays. *Annual Review of Astronomy and Astrophysics*, 22:425–444, 1984.

[HKK+88] K. S. Hirata, T. Kajita, M. Koshiba, et al. Observation in the Kamiokande-II detector of the neutrino burst from supernova SN1987A. *Phys. Rev. D*, 38(2):448–458, Jul 1988.

[Hor07] D. Horns on behalf of the HESS Collaboration. H.E.S.S.: Status and future plan. *Journal of Physics Conference Series*, 60:119–122, March 2007.

[HSG+05] J. Hjorth, J. Sollerman, J. Gorosabel, et al. GRB 050509B: Constraints on Short Gamma-Ray Burst Models. *Astrophysical Journal Letters*, 630:L117–L120, September 2005.

[I+09] J. Isbert et al. Energy Spectra from the ATIC Balloon Experiment. In *Bulletin of the American Astronomical Society*, volume 41 of *Bulletin of the American Astronomical Society*, pages 439–+, January 2009.

[IYH05] D. Iono, M. S. Yun, and P. T. P. Ho. Atomic and Molecular Gas in Colliding Galaxy Systems. I. The Data. *The Astrophysical Journal Supplement Series*, 158:1–37, May 2005.

[Jac62] John David Jackson. *Classical Electrodynamics*. John Wiley & Sons Inc., 1962.

[Jan33] Karl Guthe Jansky. Radio Waves from Outside the Solar System. *Nature*, 132:66–+, July 1933.

Bibliography

[JGDB96] J. J. Condon, G. Helou, D. B. Sanders, and B. T. Soifer. A 1.425 ghz atlas of the iras bright galaxy sample, part ii. *The Astrophysical Journal Supplement Series*, 103, 1996.

[JWE+98] J. J. Condon, W. D. Cotton, E. W. Greisen, Q. F. Yin, R. A. Perley, G. B. Taylor, and J. J. Boderick. The nrao vla sky survey. *Astronomical Journal*, 115, 1998.

[K+00] P. P. Kronberg et al. A Search for Flux Density Variations in 24 Compact Radio Sources in M82. *The Astrophysical Journal*, 535:706–711, June 2000.

[Kar08] A. Karle for the IceCube Collaboration. IceCube: Construction Status and First Results. *ArXiv e-prints: 0812.398*, December 2008.

[Ken98] R. C. Kennicutt, Jr. The Global Schmidt Law in Star-forming Galaxies. *The Astrophysical Journal*, 498:541–+, May 1998.

[Kin55] Henry C. King. *The History of the Telescope*. Charles Griffin & Co. London, 1955.

[Kna94] J. Knapp. Corrected file of data from 1989APJS...70..329K as deposited at NSSDC (J/APJS/70/329). *Private Communication*, 1994.

[KWPN81] H. Kuehr, A. Witzel, I. I. K. Pauliny-Toth, and U. Nauber. A catalogue of extragalactic radio sources having flux densities greater than 1 Jy at 5 GHz. *Astronomy and Astrophysics, Supplement*, 45:367–430, September 1981.

[Lah09] R. Lahmann. Status and first results of the acoustic detection test system AMADEUS. *Nuclear Instruments and Methods in Physics Research A*, 604:158–+, June 2009.

[LB06] B. Link and F. Burgio. Flux predictions of high-energy neutrinos from pulsars. *Monthly Notices of the Royal Astronomical Society*, 371:375–379, September 2006.

[LBSB05] A. Leroy, A. D. Bolatto, J. D. Simon, and L. Blitz. The Molecular Interstellar Medium of Dwarf Galaxies on Kiloparsec Scales: A New Survey for CO in Northern, IRAS-detected Dwarf Galaxies. *The Astrophysical Journal*, 625:763–784, June 2005.

[Lew09] T. Lewke on behalf of the Borexino Collaboration. Results from the Borexino Experiment. *ArXiv e-prints: 0905.2526*, May 2009. Proceedings for the Moriond 2009 EW session.

[Lon92] Malcom Sim Longair. *High Energy Astrophysics*, volume 1. Cambridge University Press, 1992.

[LRV08] H. Landsman, L. Ruckman, and G. S. Varner. AURA - A radio frequency extension to IceCube. *ArXiv e-prints: 0811.2520*, November 2008. To appear in the proceedings of the Acoustic and Radio EeV Neutrino detection Activities (ARENA) 2008 conference.

[LVS+07] U. Lisenfeld, L. Verdes-Montenegro, J. Sulentic, et al. The AMIGA sample of isolated galaxies. III. IRAS data and infrared diagnostics. *Astronomy & Astrophysics*, 462:507–523, February 2007.

[LW06] A. Loeb and E. Waxman. The cumulative background of high energy neutrinos from starburst galaxies. *Journal of Cosmology and Astroparticle Physics*, 5:3, 2006.

[M+03a] F. Mannucci et al. The infrared supernova rate in starburst galaxies. *Astronomy & Astrophysics*, 401:519, April 2003.

[M+03b] P. A. Mazzali et al. The Type Ic Hypernova SN 2003dh/GRB 030329. *Astrophysical Journal Letters*, 599:L95, 2003.

[M+05] D. C. Martin et al. The Galaxy Evolution Explorer: A Space Ultraviolet Survey Mission. *Astrophysical Journal Letters*, 619:L1–L6, January 2005.

[M+07] Kirsten Münich (IceCube Collaboration) et al. . In *Proceedings of the 30th International Cosmic Ray Conference ICRC, Merida, Mexico*, July 2007. ArXiv:711.0353.

[M+08] B. J. Morsony et al. Jitter radiation in gamma-ray bursts and their afterglows: emission and self-absorption. *Monthly Notices of the Royal Astronomical Society*, 386:199–210, May 2008.

[M+09] C. Meegan (Fermi GBM Collaboration) et al. The Fermi Gamma-ray Burst Monitor. *The Astrophysical Journal*, 702:791–804, September 2009.

[Mac97] Ernst Waldfried Josef Wenzel Mach. Über Erscheinungen an fliegenden Projektilen, November 1897. Talk held at 'Wiener Verein zur Verbreitung naturwissenschaftlicher Kenntnisse'.

[Man95] K. Mannheim. High-energy neutrinos from extragalactic jets. *Astroparticle Physics*, 3:295–302, May 1995.

[MDT+03] P. A. Mazzali, J. Deng, Tominaga, et al. The Type Ic Hypernova SN 2003dh/GRB 030329. *Astrophysical Journal Letters*, 599:L95–L98, December 2003.

[MG+90] M. Moshir, G. Kopan, et al. The faint source catalog, version 2.0. *Bulletin of the American Astronomical Society*, 22, 1990.

[MGI74] E. P. Mazets, S. V. Golenetskij, and V. N. Il'Inskij. Burst of cosmic gamma - emission from observations on Cosmos 461. *Pis ma Zhurnal Eksperimental noi i Teoreticheskoi Fiziki*, 19:126–128, 1974.

[MHKM09] A. McCann, D. Hanna, J. Kildea, and M. McCutcheon. A New Mirror Alignment System for the VERITAS Telescopes. *ArXiv e-prints: 0910.3277*, October 2009.

[MIC+05] Kirsten Münich (IceCube Collaboration) et al. Search for a diffuse flux of nonterrestrial muon neutrinos with the AMANDA detector. In *Proceedings of the 29th International Cosmic Ray Conference ICRC, Pune, India*, 2005. astro-ph/0509330.

[MO42] N. U. Mayall and J. H. Oort. Further Data Bearing on the Identification of the Crab Nebula with the Supernova of 1054 A.D. Part II. The Astronomical Aspects. *The Publications of the Astronomical Society of the Pacific*, 54:95–104, April 1942.

[Mon08] T. Montaruli for the IceCube Collaboration. First results of the IceCube observatory on high energy neutrino astronomy. *Journal of Physics Conference Series*, 120(6):062009–+, July 2008.

[MPE+03] A. Mücke, R. J. Protheroe, R. Engel, J. P. Rachen, and T. Stanev. BL Lac objects in the synchrotron proton blazar model. *Astroparticle Physics*, 18:593–613, March 2003.

[MPR01] K. Mannheim, R. J. Protheroe, and J. P. Rachen. Cosmic ray bound for models of extragalactic neutrino production. *Physical Review D*, 63(2):023003–+, January 2001.

[MSB92] K. Mannheim, T. Stanev, and P. L. Biermann. Neutrinos from flat-spectrum radio quasars. *Astronomy & Astrophysics*, 260:L1–L3, July 1992.

[Mün07] Kirsten Münich. *Messungen des atmosphärischen Neutrinospektrums mit dem AMANDA-II Detektor—Bestimmung eines 90% oberen Limits auf den extraterrestrischen Beitrag*. PhD thesis, Universität Dortmund, 2007.

[MW84] Ernst Mach and Josef Wentzel. Experimental setup. *Anzeiger der Kaiserlichen Akademie der Wissenschaften Wien*, 21:121, 1884.

[MW85] Ernst Mach and Josef Wentzel. Experimental results. *Anzeiger der Kaiserlichen Akademie der Wissenschaften Wien*, 92:625, 1885.

[N+09] T. Nonaka et al. The present status of the Telescope Array experiment. *Nuclear Physics B - Proceedings Supplements*, 190:26 – 31, 2009. Proceedings of the Cosmic Ray International Seminars.

[Nag04] S. Nagataki. High-Energy Neutrinos Produced by Interactions of Relativistic Protons in Shocked Pulsar Winds. *The Astrophysical Journal*, 600:883–904, January 2004.

[NFW05] N. M. Nagar, H. Falcke, and A. S. Wilson. Radio sources in low-luminosity active galactic nuclei. IV. Radio luminosity function, importance of jet power, and radio properties of the complete Palomar sample. *Astronomy & Astrophysics*, 435:521–543, May 2005.

[NMB93] L. Nellen, K. Mannheim, and P. L. Biermann. Neutrino production through hadronic cascades in AGN accretion disks. *Physical Review D*, 47:5270–5274, June 1993.

[O'D98] C. P. O'Dea. The Compact Steep-Spectrum and Gigahertz Peaked-Spectrum Radio Sources. *The Publications of the Astronomical Society of the Pacific*, 110:493–532, May 1998.

[Ott09] A. Nepomuk Otte (The VERITAS collaboration). Upgrade of the VERITAS Cherenkov Telescope Array. *ArXiv e-prints: 0907.4826*, July 2009.

[OWB05] J. Ott, F. Walter, and E. Brinks. A Chandra X-ray survey of nearby dwarf starburst galaxies - I. Data reduction and results. *Monthly Notices of the Royal Astronomical Society*, 358:1423–1452, April 2005.

[PMP+99] W. S. Paciesas, C. A. Meegan, G. N. Pendleton, M. S. Briggs, C. Kouveliotou, T. M. Koshut, J. P. Lestrade, M. L. McCollough, J. J. Brainerd, J. Hakkila, W. Henze, R. D. Preece, V. Connaughton, R. M. Kippen, R. S. Mallozzi, G. J. Fishman, G. A. Richardson, and M. Sahi. The Fourth BATSE Gamma-Ray Burst Catalog (Revised). *The Astrophysical Journal Supplement Series*, 122:465–495, June 1999.

[Pro98] Raymond J. Protheroe. Origin and Propagation of the Highest Energy Cosmic Rays. In M. M. Shapiro, R. Silberberg, & J. P. Wefel, editor, *Towards the Millennium in Astrophysics, Problems and Prospects. International School of Cosmic Ray Astrophysics 10th Course*, pages 3–+, 1998.

[Pro04] Raymond J. Protheroe. Effect of energy losses and interactions during diffusive shock acceleration: applications to SNR, AGN and UHE cosmic rays. *Astroparticle Physics*, 21:415–431, July 2004.

[R+93] P. M. Rodriguez-Pascual et al. X-ray, UV and FIR emission of Seyfert galaxies. *Astrophysics and Space Science*, 205:113, July 1993.

[R+08] J. L. Racusin et al. Broadband observations of the naked-eye γ-ray burst GRB080319B. *Nature*, 455:183, September 2008.

[RA08] J. M. Rodriguez Espinosa and P. Alvarez Martin. The GTC facility instruments: a status review. In *Society of Photo-Optical Instrumentation Engineers (SPIE) Conference Series*, volume 7014 of *Presented at the Society of Photo-Optical Instrumentation Engineers (SPIE) Conference*, August 2008.

[Ran09] R. Rando on behalf of the Fermi LAT Collaboration. Post-launch performance of the Fermi Large Area Telescope. *ArXiv e-prints: 0907.0626*, July 2009. Proceedings of the 31st ICRC Łódź, Poland.

[Rap08] P. A. Rapidis. KM3NeT: a large underwater neutrino telescope in the Mediterranean sea. *Journal of Physics Conference Series*, 120(6):062011–+, July 2008.

[RC+56] Frederick Reines, Clyde L. Cowan, et al. Detection of the free neutrino: A confirmation. *Science*, (124):103, 1956.

[Reb44] Grote Reber. Cosmic Static. *The Astrophysical Journal*, 100:279–+, November 1944.

[RFG98] M. W. Richmond, A. V. Filippenko, and J. Galisky. The Supernova Rate in Starburst Galaxies. *The Publications of the Astronomical Society of the Pacific*, 110:553, May 1998.

[RLR96] D. Rigopoulou, A. Lawrence, and Rowan-Robinson. Multiwavelength energy distributions of ultraluminous IRAS galaxies - I. Submillimetre and X-ray observations. *Monthly Notices of the Royal Astronomical Society*, 278:1049–1068, February 1996.

[RMW03] S. Razzaque, P. Mészáros, and E. Waxman. Neutrino tomography of gamma ray bursts and massive stellar collapses. *Physical Review D*, 68(8):083001–+, October 2003.

[Rou00] Esteban Roulet. Neutrinos in Physics and Astrophysics. In D. Page & J. G. Hirsch, editor, *From the Sun to the Great Attractor*, volume 556 of *Lecture Notes in Physics, Berlin Springer Verlag*, pages 233–+, 2000.

[RR92] Richard L. White and Robert H. Becker. A new catalog of 30,239 1.4 ghz sources. *The Astrophysical Journal Supplement Series*, 79, 1992.

[RRA91] Robert H. Becker, Richard L. White, and Alan L. Edwards. A new catalog of 53,522 4.85 ghz sources. *The Astrophysical Journal Supplement Series*, 75, 1991.

[RRD95] Robert H. Becker, Richard L. White, and David J. Helfland. The first survey: Faint images of the radio sky at twenty centimeters. *The Astrophysical Journal*, 450, 1995.

[RSTT07] D. Rosa-González, H. R. Schmitt, E. Terlevich, and R. Terlevich. Thermal Emission from H II Galaxies: Discovering the Youngest Systems. *The Astrophysical Journal*, 654:226–239, January 2007.

[S+07] X. W. Shu et al. Investigating the Nuclear Obscuration in Two Types of Seyfert 2 Galaxies. *The Astrophysical Journal*, 657:167–176, March 2007.

[S+09] V. de Souza (KASCADE-Grande Collaboration) et al. The KASCADE-Grande Experiment. In C. J. Solano Salinas, J. Bellido, D. Wahl, & O. Saavedra, editor, *American Institute of Physics Conference Series*, volume 1123 of *American Institute of Physics Conference Series*, pages 211–216, April 2009.

[Sat77] H. Sato. Pulsars Covered by the Dense Envelopes as High-Energy Neutrino Sources. *Progress of Theoretical Physics*, 58:549–559, August 1977.

[SBG93] T. Stanev, P. L. Biermann, and T. K. Gaisser. Cosmic rays. IV. The spectrum and chemical composition above 10' GeV. *Astronomy & Astrophysics*, 274:902–+, July 1993.

[SBNS89] B. T. Soifer, L. Boehmer, G. Neugebauer, and D. B. Sanders. The IRAS Bright Galaxy Sample. IV - Complete IRAS observations. *Astronomical Journal*, 98:766–797, September 1989.

[Sch59] M. Schmidt. The Rate of Star Formation. *The Astrophysical Journal*, 129:243–+, March 1959.

[Sch63]   M. Schmidt. 3C 273 : A Star-Like Object with Large Red-Shift. *Nature*, 197:1040–+, March 1963.

[Sch97]   Norbert Schmitz. *Neutrinophysik*. Teubner-Studienbücher, 1997.

[SGW+01] D. Saltzberg, P. Gorham, D. Walz, et al. Observation of the Askaryan Effect: Coherent Microwave Cherenkov Emission from Charge Asymmetry in High-Energy Particle Cascades. *Physical Review Letters*, 86:2802–2805, March 2001.

[SHC+04] D. K. Strickland, T. M. Heckman, E. J. M. Colbert, C. G. Hoopes, and K. A. Weaver. A High Spatial Resolution X-Ray and Hα Study of Hot Gas in the Halos of Star-forming Disk Galaxies. II. Quantifying Supernova Feedback. *The Astrophysical Journal*, 606:829–852, May 2004.

[SKO73]   I. B. Strong, R. W. Klebesadel, and R. A. Olson. Possible Detection of Prompt Gamma Rays from Supernovae. In *Bulletin of the American Astronomical Society*, volume 5 of *Bulletin of the American Astronomical Society*, pages 322–+, June 1973.

[SMAD85] H. Spinrad, J. Marr, L. Aguilar, and S. Djorgovski. A third update of the status of the 3CR sources - Further new redshifts and new identifications of distant galaxies. *The Publications of the Astronomical Society of the Pacific*, 97:932–961, October 1985.

[SMK+03] D. B. Sanders, J. M. Mazzarella, D.-C. Kim, J. A. Surace, and B. T. Soifer. The IRAS Revised Bright Galaxy Sample. *Astronomical Journal*, 126:1607–1664, October 2003.

[SNO02]   SNO-Kollaboration. Direct Evidence for Neutrino Flavor Transformation from Neutral-Current Interactions in the Sudbury Neutrino Observatory. *Physical Review Letters*, 89:11301, 2002.

[Sra75]   R. Sramek. 5-GHz survey of bright galaxies. *Astronomical Journal*, 80:771–777, October 1975.

[SS96]   F. W. Stecker and M. H. Salamon. High Energy Neutrinos from Quasars. *Space Science Reviews*, 75:341–355, January 1996.

[SSM04]   J. A. Surace, D. B. Sanders, and J. M. Mazzarella. An IRAS High Resolution Image Restoration (HIRES) Atlas of All Interacting Galaxies in the IRAS Revised Bright Galaxy Sample. *Astronomical Journal*, 127:3235–3272, June 2004.

[SSWS09] G. K. Squires, L. J. Storrie-Lombardi, M. W. Werner, and B. T. Soifer. Spitzer: Hot Science in the Warm Mission. In *American Astronomical Society, AAS Meeting #214, #202.01*, volume 214 of *Bulletin of the American Astronomical Society*, page 710, May 2009.

[ST76]   R. A. Sramek and H. M. Tovmassian. The radio brightness distribution of eight Markarian galaxies. *The Astrophysical Journal*, 207:725–729, August 1976.

Bibliography

[Sta04]   Todor Stanev. *High Energy Cosmic Rays*. Springer, second edition, 2004.

[Sta09]   T. Stanev for the IceCube Collaboration. Status, performance, and first results of the IceTop array. *Nuclear Physics B Proceedings Supplements*, 196:159–164, December 2009.

[Ste05]   F. W. Stecker. Note on high-energy neutrinos from active galactic nuclei cores. *Physical Review D*, 72(10):107301–+, November 2005.

[Ste07]   F. W. Stecker. Are diffuse high energy neutrinos and γ-rays from starburst galaxies observable? *Astroparticle Physics*, 26:398, 2007.

[SW09]   L. A. Sargsyan and D. W. Weedman. Star Formation Rates for Starburst Galaxies from Ultraviolet, Infrared, and Radio Luminosities. *The Astrophysical Journal*, 701:1398–1414, August 2009.

[T+78]   G. I. Thompson et al. Catalogue of stellar ultraviolet fluxes. A compilation of absolute stellar fluxes measured by the Sky Survey Telescope (S2/68) aboard the ESRO satellite TD-1. 1978.

[T+06]   T. A. Thompson et al. Magnetic Fields in Starburst Galaxies and the Origin of the FIR-Radio Correlation. *The Astrophysical Journal*, 645:186, July 2006.

[The09]   The Veritas Collaboration. VERITAS Collaboration Contributions to the 31st International Cosmic Ray Conference. *ArXiv e-prints: 0908.0150*, August 2009.

[TPR+06]   R. Tüllmann, W. Pietsch, J. Rossa, D. Breitschwerdt, and R.-J. Dettmar. The multi-phase gaseous halos of star forming late-type galaxies. I. XMM-Newton observations of the hot ionized medium. *Astronomy & Astrophysics*, 448:43–75, March 2006.

[TTW+05]   M. Tajer, G. Trinchieri, A. Wolter, S. Campana, A. Moretti, and G. Tagliaferri. A sample of X-ray emitting normal galaxies from the BMW-HRI Catalogue. *Astronomy & Astrophysics*, 435:799–810, June 2005.

[TWV+05]   S. H. Teng, A. S. Wilson, Veilleux, et al. A Chandra X-Ray Survey of Ultraluminous Infrared Galaxies. *The Astrophysical Journal*, 633:664–679, November 2005.

[Ulv00]   James S. Ulvestad. Circumnuclear Supernova Remnants and H II Regions in NGC 253. *Astronomical Journal*, 120:278–283, July 2000.

[VAB96]   H. J. Völk, F. A. Aharonian, and D. Breitschwerdt. The Nonthermal Energy Content and Gamma-Ray Emission of Starburst Galaxies and Clusters of Galaxies. *Space Science Reviews*, 75:279–297, January 1996.

[VBD08]   B. Vollmer, T. Beckert, and R. I. Davies. Starbursts and torus evolution in AGN. *Astronomy & Astrophysics*, 491:441–453, November 2008.

[VGL05] J. Vandenbroucke, G. Gratta, and N. Lehtinen. Experimental Study of Acoustic Ultra-High-Energy Neutrino Detection. *The Astrophysical Journal*, 621:301–312, March 2005.

[Vie95] M. Vietri. The Acceleration of Ultra–High-Energy Cosmic Rays in Gamma-Ray Bursts. *The Astrophysical Journal*, 453:883, November 1995.

[VLR+05] J. S. Villasenor, D. Q. Lamb, G. R. Ricker, et al. Discovery of the short γ-ray burst GRB 050709. *Nature*, 437:855–858, October 2005.

[Vol80] L. V. Volkova. Energy spectra and angular distribution of atmospheric neutrinos. *Sov. J. of Nucl. Phys.*, 31:748, 1980.

[VTW04] B. Vollmer, M. Thierbach, and R. Wielebinski. Radio continuum spectra of galaxies in the Virgo cluster region. *Astronomy & Astrophysics*, 418:1–6, April 2004.

[VZ80] Lyudmilla Valeriyevna Volkova and Georgi Timofeevič Zatsepin. Promt lepton generation-atmospheric muon and neutrino spectra at high-energies. *Soviet Journal of Nuclear Physics*, 37:212, 1980.

[Wag04] Wolfgang Wagner. *Design and Realisation of a new AMANDA Data Acquisition System with Transient Waveform Recordes*. PhD thesis, Universität Dortmund, October 2004. DissDo 2004-169.

[Wax95] E. Waxman. Cosmological Gamma-Ray Bursts and the Highest Energy Cosmic Rays. *Physical Review Letters*, 75:386, July 1995.

[WB97] E. Waxman and J. N. Bahcall. High Energy Neutrinos from Cosmological Gamma-Ray Burst Fireballs. *Physical Review Letters*, 78:2292, 1997.

[WB99] E. Waxman and J. Bahcall. High energy neutrinos from astrophysical sources: An upper bound. *Physical Review D*, 59(2):023002–+, January 1999.

[WB00] E. Waxman and J. N. Bahcall. Neutrino Afterglow from Gamma-Ray Bursts: $\sim 10^{18}$ eV. *The Astrophysical Journal*, 541:707–711, October 2000.

[WBC+02] M. C. Weisskopf, B. Brinkman, C. Canizares, G. Garmire, S. Murray, and L. P. Van Speybroeck. An Overview of the Performance and Scientific Results from the Chandra X-Ray Observatory. *The Publications of the Astronomical Society of the Pacific*, 114:1–24, January 2002.

[WGA00] N. E. White, P. Giommi, and L. Angelini. The WGACAT version of ROSAT sources (White+ 2000). *VizieR Online Data Catalog*, 9031:0–+, June 2000.

[WGBE94] A. E. Wright, M. R. Griffith, B. F. Burke, and R. D. Ekers. The Parkes-MIT-NRAO (PMN) surveys. 2: Source catalog for the southern survey ($-87.5 < \delta < -37°$). *The Astrophysical Journal Supplement Series*, 91:111–308, March 1994.

[WGC+94] S. L. Wheelock, T. N. Gautier, Chillemi, et al. IRAS sky survey atlas: Explanatory supplement. *NASA STI/Recon Technical Report N*, 95:22539–+, May 1994.

[WGH+96] A. E. Wright, M. R. Griffith, A. J. Hunt, E. Troup, B. F. Burke, and R. D. Ekers. The Parkes-MIT-NRAO (PMN) Surveys. VIII. Source Catalog for the Zenith Survey ($-37.0° < \delta < -29.0°$). *The Astrophysical Journal Supplement Series*, 103:145–+, March 1996.

[Wie98] Barbara Wiebel Sooth. *Measurement of the allparticle energy spectrum and chemical composition of cosmic rays with the HEGRA detector*. PhD thesis, Universität Wuppertal, September 1998.

[Wie09] C. Wiebusch for the IceCube Collaboration. Physics Capabilities of the IceCube DeepCore Detector. *ArXiv e-prints: 0907.2263*, July 2009. Proceedings of the 31st ICRC Łódź, Poland.

[Wil01] Charles Thompson Rees Wilson. On the Ionisation of Atmospheric Air. *Proceedings of the Royal Society of London*, 68:151–161, January 1901.

[WRM97] R. A. M. J. Wijers, M. J. Rees, and P. Meszaros. Shocked by GRB 970228: the afterglow of a cosmological fireball. *Monthly Notices of the Royal Astronomical Society*, 288:L51–L56, July 1997.

[WUB+73] W. A. Wheaton, M. P. Ulmer, W. A. Baity, D. W. Datlowe, M. J. Elcan, L. E. Peterson, R. W. Klebesadel, I. B. Strong, T. L. Cline, and U. D. Desai. The Direction and Spectral Variability of a Cosmic Gamma-Ray Burst. *Astrophysical Journal Letters*, 185:L57+, October 1973.

[X+94] C. Xu et al. Star-formation histories and the mass-normalized FIR/ratio correlation in late-type galaxies. *Astronomy & Astrophysics*, 282:19, February 1994.

[YT93] S. Yoshida and M. Teshima. Energy Spectrum of Ultra-High Energy Cosmic Rays with Extra-Galactic Origin. *Progress of Theoretical Physics*, 89:833–845, April 1993.

[ZB02] C. Zier and P. L. Biermann. Binary black holes and tori in AGN. II. Can stellar winds constitute a dusty torus? *Astronomy & Astrophysics*, 396:91–108, December 2002.

[ZK66] Georgi Timofeevič Zatsepin and Vadim Alexeevič Kuzmin. Upper limit of the spectrum of cosmic rays. *Pisma Zhurnal Eksp. Teor. Fiziki*, 4:114, 1966.

Die VDM Verlagsservicegesellschaft sucht für wissenschaftliche Verlage abgeschlossene und herausragende

Dissertationen, Habilitationen, Diplomarbeiten, Master Theses, Magisterarbeiten usw.

für die kostenlose Publikation als Fachbuch.

Sie verfügen über eine Arbeit, die hohen inhaltlichen und formalen Ansprüchen genügt, und haben Interesse an einer honorarvergüteten Publikation?

Dann senden Sie bitte erste Informationen über sich und Ihre Arbeit per Email an *info@vdm-vsg.de*.

Sie erhalten kurzfristig unser Feedback!

VDM Verlagsservicegesellschaft mbH
Dudweiler Landstr. 99
D - 66123 Saarbrücken
www.vdm-vsg.de

Telefon +49 681 3720 174
Fax +49 681 3720 1749

Die VDM Verlagsservicegesellschaft mbH vertritt

Printed by Books on Demand GmbH, Norderstedt / Germany